出得廳堂宴客菜

Delicacies for your Guests

編者話

Preface

每年，總有幾次團圓歡聚的機會——團年飯、開年飯、中秋節、冬至、父母親節、生日飯等，在酒樓吃膩了千篇一律的菜式，何不自組設計餐單歡宴親友更顯誠意？

想吃得豪華，可選～豬手凍、鮮菌石斑球、芝士南瓜汁龍蝦球、四寶釀子雞、蟹粉獅子頭、瑤柱汁燴鮮鮑……

想吃得健康一點，可選～腐皮千層、碧綠桂魚片、蟹肉桂花翅、菜膽上湯雞、杏鮑菇炒牛柳、海膽菜粒炒飯……

對炮製宴席沒有信心嗎？不要緊，食譜內詳列多項技巧，如魚肉切片及切球、龍蝦劏製及起肉、製作冰豆腐等，圖文並茂，依着指示逐步烹調，一席出得廳堂的宴客菜式，肯定令賓客開懷、盡興！

In Hong Kong, family and good friends meet at least a few times a year – around Chinese New Year, mid-Autumn, winter solstice, Mother's and Father's Day, birthdays etc. Restaurants serve the same dishes and you may get sick of them after a few times. Why don't you create a menu specifically for your guests to show your love and care?

For a luxurious gourmet feast, start with a pork trotter aspic, followed by braised grouper with fresh mushrooms, lobster chunks with cheese and pumpkin sauce, steamed chicken stuffed with four delicacies, crab roe and pork balls with white cabbage, braised abalones with dried scallops extract...

For a healthy meal, make steamed tofu skin roll, stir-fried mandarin fish with broccoli, scrambled eggs with shark's fin and mung bean sprouts, steamed chicken in stock with mustard green, stir-fried beef tenderloin with king oyster mushrooms, stir fried rice with sea urchin and vegetable...

For kitchen newbies not confident in hosting a party, no worries. This cookbook details every cooking skill, such as slicing fish, dressing and shelling lobster, and making frozen tofu. All recipes are fully illustrated. As long as you follow the steps closely, everyone can churn out a table full of sumptuous and presentable dishes. I'm sure your guests will enjoy themselves to the fullest.

目錄
Contents

主菜・蝦蟹
Main Dish・Crustacean

胡椒蝦 / 62
Fried Prawns with
White Peppercorns

薑葱炒肉蟹 / 64
Stir-fried Male Mud Crab
with Ginger and Spring Onion

蟹粉獅子頭 / 67
Crab Roe and Pork Balls
with White Cabbage

芝士南瓜汁龍蝦球 / 70
Lobster Chunks with
Cheese and Pumpkin Sauce

蟹肉桂花翅 / 74
Scrambled Eggs with Shark's Fin
and Mung Bean Sprouts

南瓜黃金蝦 / 77
Stir-fried Prawns with
Pumpkin and Salted Egg Yolk

乾葱豉油焗中蝦 / 80
Fried Medium Prawns with
Shallot and Light Soy Sauce

主菜・貝類
Main Dish・Shellfish

鮑汁金銀帶子 / 82
Fried Scallops with
Abalone Sauce and
Dried Scallop Shreds

香草牛油焗翡翠螺 / 85
Baked Green Whelks with
Herbs and Butter

XO 醬煎金蠔 / 88
Fried Golden Dried Oysters
with XO Sauce

蛋白蒸海膽 / 90
Steamed Sea Urchin and
Egg Whites

瑤柱汁燴鮮鮑 / 92
Braised Abalones with
Dried Scallops Extract

背煎豉汁蟶子皇 / 95
Fried Razor Clams in
Black Bean Sauce

帶子百花石榴粿 / 98
Steamed Scallops and
Prawns Beggar's Purses

飯麵・甜品
Staple・Desserts

海膽菜粒炒飯 / 102
Stir-fried Rice with Sea Urchin
and Vegetable

黃金豆腐醬伴脆米粉 / 104
Crispy Rice Vermicelli with Salted
Egg Yolk and Tofu Paste

芝士焦糖脆蛋 / 107
Cheese Crème Brûlée

豆腐西米焗布甸 / 110
Baked Tofu and
Sago Pudding

做一客宴席攻略
Tips on Hosting a Dinner Party

宴請友人回家飯聚，作為主人家，每一個細節都要細心思量，無論在預備美食、佈置裝飾及其他方面，都為賓客設想周到，盡興而回！

預備菜單：

- 向賓客詢問有否食物過敏病史，盡量避免引起致敏的食材，例如雞蛋、魚、蝦蟹或奶類等，可嘗試以其他食材代替。
- 若賓客有老人家或小朋友，建議魚肉去骨起肉，以軟腍的冬瓜、南瓜、帶子、蒸蛋白等作為餐單，方便享用；另外，小朋友應避免進食用酒醃製的菜式，如醉雞。
- 預早清楚知道那位賓客是茹素者，為他們安排特別的餐單菜式吧！
- 現今講求健康飲食，味道別太濃重，緊記鹹淡相宜，品嘗濃味菜式後，可配搭一款清新的食品，有一種洗滌味蕾的感覺。
- 建議別偏重某類食材，海鮮、肉類、家禽、蔬菜、菇菌等排入你的菜單內，豐富宴席內容。
- 如時間許可，安排飯後甜品，為宴席完結前留下一道甜美的回憶。

預先準備：

- 冷盤食物可預先製作妥當，冷吃的取出即可食用；熱吃的上菜前翻熱或炒煮皆可。
- 蝦蟹、魚及龍蝦等，不宜太早處理或劏製，以免鮮味流失。
- 需要浸發的食材，如冬菇、乾瑤柱、竹笙等，宜早一天浸發妥當，並冷藏備用。
- 需長時間醃製的全雞或肉類，拌勻醃料後，宜放雪櫃待一晚，省卻宴席當天的烹調時間。
- 自製的醬汁或調味料，適宜預早一兩天備好，烹調時更得心應手。
- 可使用多種類的烹調用具，如焗爐、蒸鍋、真空煲等，可同一時間烹煮多款菜式，節省時間。

As a host, you need to carefully think about every detail, from food preparations to decorating your dining room. The aim is to include every guest mindfully in the process, so that everyone has a good time.

Drafting a menu

- Ask your guests if they have any allergy. Try your best to avoid the most common foods that cause allergies, such as eggs, fish, crustaceans and dairy products. Replace them with other ingredients if possible.
- If there are young children or senior members among your guests, debone all fish. Include softer ingredients in your menu, such as winter melon, pumpkin, scallops or steamed custard. Also avoid dishes that use alcohol generously, such as drunken chicken, if there are young children.
- Check if there's any vegetarian among your guests. Tailor-make a menu for him/her.
- People are health-conscious these days. Try not to over-season your food. After a rich-tasting dish, it's advisable to serve a light dish to cleanse the palate.
- Try not to bias on a certain type of ingredients. Strike a balance among seafood, meat, poultry, vegetables and mushrooms. Instead of a one-note meal, create a varied dining experience.
- If time allows, serve dessert to end the meal in the perfect way.

Preparations

- Make all appetizers in advance. Serve cold ones straight out of the fridge. For hot ones, reheat them or stir-fry them till hot before serving.
- For seafood such as live crabs, shrimps, fish or lobsters, do not slaughter or dress them too early in advance. Always dress them right before you cook them. Otherwise, they won't taste as good and fresh.
- Dried ingredients that need rehydration, such as shiitake mushrooms, dried scallops, and bamboo fungus, should be soak in water the day before the dinner party. Keep them in the fridge until ready to cook.
- Meat or whole chicken takes a long time to marinate. Thus, you should mix them well in the marinade one day before the dinner party to save time.
- Make the sauces or seasoning one or two days in advance. You can manage the time much better on the cooking day.
- Try to design a menu that involves different cooking tools so that several dishes can be made at the same time, for instance, an oven, a steamer, and a pressure cooker. That would save you lots of time.

吃得健康

Dinner Menu for the Health-conscious

在追求健康飲食的年代，部份人捨棄大魚大肉的菜式，轉而追求簡單健營的食物。在設計餐單時，可考慮以下菜式建議，為你賓客的健康加點力！

類別	菜式	特點	參考頁
冷盤	腐皮千層	包含菇菌蔬菜，全素食品，蒸吃	p.10
	桂花酒醉雞	蒸熟後用桂花陳酒浸醃	p.14
主菜・肉類	洋葱炒黑毛豬腩片	黑毛豬腩片肥瘦均稱	p.28
	杏鮑菇炒牛柳	杏鮑菇高纖維，降血脂及膽固醇	p.32
主菜・家禽	菜膽上湯雞	蒸熟伴菜膽上桌，少調味	p.36
	酸梅鴨	鴨件用焗爐烤焗，迫出油分	p.44
主菜・魚類	鮮菌石斑球	材料是斑肉、菇菌及蔬菜，炒熟即成	p.46
	冬瓜金腿魚夾	冬瓜伴斑肉，同時吸收纖維	p.52
主菜・蝦蟹	芝士南瓜汁龍蝦球	選用有營的南瓜煮汁伴吃	p.70
	蟹肉桂花翅	調味量少，快炒	p.74
主菜・貝類	蛋白蒸海膽	蛋白健康，少調味	p.90
	瑤柱汁燴鮮鮑	鮑魚蘊含豐富的蛋白質	p.92
飯麵・甜品	海膽菜粒炒飯	攝取纖維及蛋白質，少調味	p.102
	豆腐西米焗布甸	吸收豆腐的植物性蛋白質	p.110

Healthy diet is all the rage. Some people are switching from heavy meals to a light but nutritious diet. When you design your menu for a dinner party, you may consider making the followings for those health conscious.

Course	Recipe	Characteristics	page
Appetizer	Steamed Tofu Skin Roll	It includes mushrooms and veggies. Steam it and serve.	p.10
	Drunken Chicken in Osmanthus Wine	Steam the chicken and steep it in a wine marinade.	p.14
Main dish / meat	Stir-fried Kurobuta Pork Belly with Onion	Kurobuta pork belly has even marbling of fat and meat.	p.28
	Stir-fried Beef Tenderloin with King Oyster Mushrooms	King oyster mushroom is high in dietary fibre. It helps reduce blood triglyceride and cholesterol levels.	p.32
Main dish / poultry	Steamed Chicken in Stock with Mustard Green	The chicken is steamed and served with leafy greens on the side. The chicken is sparingly seasoned.	p.36
	Braised Duck in Plum Sauce	The duck is grilled in an oven so that excessive grease is drained in the cooking process.	p.44
Main dish / fish	Braised Grouper with Fresh Mushrooms	It is mostly made up of grouper fillet, mushrooms and veggies. Just stir-fry them till done.	p.46
	Steamed Winter Melon, Grouper and Jinhua Ham Sandwiches	Grouper fillet is paired with winter melon for additional dietary fibre.	p.52
Main dish / crustacean	Lobster Chunks with Cheese and Pumpkin Sauce	The nutritious pumpkin sauce makes the dish more healthful.	p.70
	Scrambled Eggs with Shark's Fin and Mung Bean Sprouts	It is a quick recipe with little seasoning.	p.74
Main dish/ shellfish	Steamed Sea Urchin and Egg Whites	Egg white is a great source of protein without cholesterol. This dish is sparingly seasoned.	p.90
	Braised Abalones with Dried Scallops Extract	Abalone is rich in protein.	p.92
Staple / desserts	Stir-fried Rice with Sea Urchin and Vegetable	The dish is sparingly seasoned, with much dietary fibre and protein.	p.102
	Baked Tofu and Sago Pudding	Tofu is a good source of botanical protein.	p.110

腐皮千層

Steamed Tofu Skin Roll

材料（6 人份量）

鮮腐皮 4 張
乾冬菇 8 朵
雲耳 2 湯匙
紅蘿蔔絲 1 碗
薑絲半湯匙

調味料

蠔油 1.5 湯匙
生抽 2 茶匙
麻油、糖各 1 茶匙
水 4 湯匙

預先準備

1. 冬菇去蒂，用水浸軟，洗淨，切絲。
2. 雲耳用水浸軟，去硬蒂，飛水，過冷河，切絲。

做法

1. 燒熱鑊，下油 1 湯匙，炒香薑絲，加入冬菇、雲耳及紅蘿蔔絲炒勻，下調味料用中小火炒至汁液濃稠，即成餡料，待涼。
2. 剪去腐皮的硬邊，每張剪成兩塊，修剪成長方形腐皮，共 4 塊。（圖 1）
3. 將 2 塊腐皮重疊鋪平，於中央位置均勻地排上一層餡料，將左右兩側之腐皮向內對摺至中央，再於一側腐皮上排滿餡料，對摺，用牙籤固定兩端，掃上餡料汁液。（圖 2-7）
4. 隔水用中火蒸 10 分鐘，取出待涼，切塊食用。

零失敗技巧
Successful cooking skills

腐皮千層可提早預備嗎？

絕對可以，包妥後可冷藏一天才蒸；或蒸後可冷藏 2 天後享用，方便宴席時使用。

Can we prepare this dish early?

Yes, of course. You can wrap and freeze it one day before steaming, or serve after steaming and freezing for two days.

為何鋪上兩層腐皮？

以免腐皮太薄餡料容易穿透，而且重疊兩張腐皮可形成多層次的效果，賣相美觀。

Why use two layers of tofu skin?

It is to prevent the filling from breaking open the thin skin, and to make the skin appear in multi-layers.

包裹腐皮有何技巧？

餡料必須待至涼透；另外，汁料及餡料不可太多，以免弄破腐皮。

What is the tips for wrapping with tofu skin?

Cool the filling thoroughly. Do not put in too much sauce and filling; otherwise the skin will break.

1
2
3
4
5
6
7

Ingredients (Serves 6)

4 sheets fresh tofu skin
8 dried black mushrooms
2 tbsp dried cloud ear fungus
1 bowl shredded carrot
1/2 tbsp shredded ginger

Seasoning

1.5 tbsp oyster sauce
2 tsp light soy sauce
1 tsp sesame oil
1 tsp sugar
4 tbsp water

Preparation

1. Remove the stalks of the dried black mushrooms. Soak in water until soft. Rinse and shred.
2. Soak the dried cloud ear fungus in water until soft. Remove the stalks. Blanch and rinse with the cold water. Shred.

Method

1. Heat a wok. Put in 1 tbsp of oil. Stir-fry the ginger until fragrant. Add the mushrooms, cloud ear fungus and carrot. Stir-fry and mix well. Put in the seasoning. Stir-fry over medium-low heat until the sauce reduces. Set aside the filling to cool.
2. Cut away the hard edge of the tofu skin. Trim each skin into 2 rectangular pieces making a total of 4. (picture 1)
3. Lay flat 2 pieces of skin overlapping each other. Evenly arrange a layer of the filling in the middle of the skin. Fold the left and right sides of the skin inwards to the middle. Overlay the filling on one side of the skin in full. Fold the skin. Fix the two ends with toothpicks. Brush the cooking sauce of the filling on the skin. (pictures 2-7)
4. Steam over medium heat for 10 minutes. Set aside to cool. Cut into pieces and serve.

桂花酒醉雞

Drunken Chicken in Osmanthus Wine

材料（6 人份量）

光雞 1 隻（約 2 斤至 2.5 斤）
桂花陳酒 1 杯（250 毫升）
桂花 1.5 湯匙

醃料

鹽半湯匙
糖 1/4 茶匙
桂花陳酒 1.5 湯匙

調味料

清雞湯 3 杯（750 毫升）
蒸雞汁 2 1/4 杯
鹽 1 湯匙
糖 3/4 茶匙

預先準備

光雞洗淨、抹乾，將醃料（桂花陳酒除外）拌勻塗抹雞皮及雞腔，再塗上桂花陳酒醃半小時。

做法

1. 燒滾水，放入雞蒸半小時至全熟，待涼，斬成四件；蒸雞汁留用。
2. 煮滾調味料，加入桂花滾 5 分鐘，汁料涼透後，加入桂花陳酒拌成浸雞汁。
3. 雞件放入步驟（2）的汁料中浸泡，冷藏一夜，取出，斬件，澆上適量桂花浸汁即可享用。

Ingredients (Serves 6)

1 chicken (about 1.2 kg to 1.5 kg)
1 cup osmanthus wine (250 ml)
1.5 tbsp dried osmanthus

Marinade

1/2 tbsp salt
1/4 tsp sugar
1.5 tbsp osmanthus wine

Seasoning

3 cups chicken stock (750 ml)
2 1/4 cups sauce from steaming chicken
1 tbsp salt
3/4 tsp sugar

Preparation

Rinse the chicken. Wipe dry. Mix the marinade (except the osmanthus wine) and spread on the chicken outside and inside. Spread the osmanthus wine on the chicken and rest for 30 minutes.

Method

1. Bring water to the boil. Steam the chicken over the water for 30 minutes until thoroughly cooked. Let it cool. Chop into 4 pieces. Reserve the sauce for later use.
2. Bring the seasoning to the boil. Add the dried osmanthus and blanch for 5 minutes. When it cools, mix in the osmanthus wine to become the sauce for soaking the chicken.
3. Soak the chicken in the sauce from step (2). Refrigerate overnight. Remove and chop into pieces. Pour some sauce over the chicken when serving.

零失敗技巧
Successful cooking skills

買不到桂花可省略嗎？

桂花於茶行容易購買；若買不到合意的桂花可省略，但浸雞汁卻少了桂花香氣，略為失色！

Can we skip dried osmanthus if they are not available?

Dried osmanthus are usually available at tea house. You can skip it if there are no suitable choices. However, it is less delicious without their fragrance in the sauce.

桂花陳酒容易購買嗎？

於普通大型超市有售，價錢相宜，帶濃濃的桂花香氣！

Is it easy to buy osmanthus wine?

With a rich osmanthus aroma, the wine can be found in supermarkets at reasonable prices.

若蒸雞汁不足份量，怎辦？

可加添清水調勻即可。

What to do if there is not enough sauce from steaming the chicken?

Add water and mix well.

豬手凍

Pork Knuckle in Jelly

材料（6 人份量）

急凍元蹄1隻（約900克）
雞腳 4 隻
冰糖 1 湯匙
紹酒 1 湯匙
鹽 2 茶匙

滷水料

薑、香葉各 2 片
八角 2 粒
花椒 1 茶匙
桂皮 1 小片
草果 1 個（拍散）

蘸汁（拌勻）

白醋 2 湯匙
蒜茸 1 茶匙
紅椒粒少許

預先準備

1. 元蹄放於雪櫃下層自然解凍，斬件，與雞腳煮約 10 分鐘，過冷河，盛起。
2. 滷水料放入魚袋內，備用。

做法

1. 滷水料用水 6 杯煮滾，加入元蹄件、雞腳、冰糖及紹酒煮約 2 小時至肉質酥軟，下鹽續煮片刻，盛起。預留燜元蹄湯汁 1.5 杯，撇去油分。
2. 元蹄待涼，去骨，撕出元蹄肉，外皮剪碎，放入玻璃容器內。（圖 1-3）
3. 元蹄湯汁隔渣，傾入容器內，冷藏約 4 小時至凝固。（圖 4）
4. 豬手凍取出，切件，伴蘸汁食用。

零失敗技巧
Successful cooking skills

將元蹄飛水及過冷河，口感如何？
可去除冷藏味，而且皮爽肉滑！
Why did you blanch the pork trotters in boiling water and rinse them in cold water?
This step helps remove the stale taste common in frozen food. It also crisps up the skin while keeping the flesh juicy.

放入雞腳燜煮有何作用？
雞腳膠質重，搭配元蹄同燜，令湯汁有大量的膠質元素，容易凝成豬手凍。
Why cook with chicken feet?
Chicken feet are rich in collagen. The cooking sauce will contain abundant collagen constituents by stewing them with pork knuckle, which is easily set into jelly.

吃起來如何有口感？
元蹄起肉時，建議撕成大塊，冷凍後口感豐富。
How to make it chewy?
Tear the meat off the pork knuckle into large pieces. It tastes complicated in cold.

這道餸可作為宴客菜嗎？
當然可以！作為冷盤小吃，大方得體，而且賓客可感受到你的誠意。
Can the dish be served at banquets?
Yes, of course! It is a beautiful cold starter and your guests will feel your sincerity.

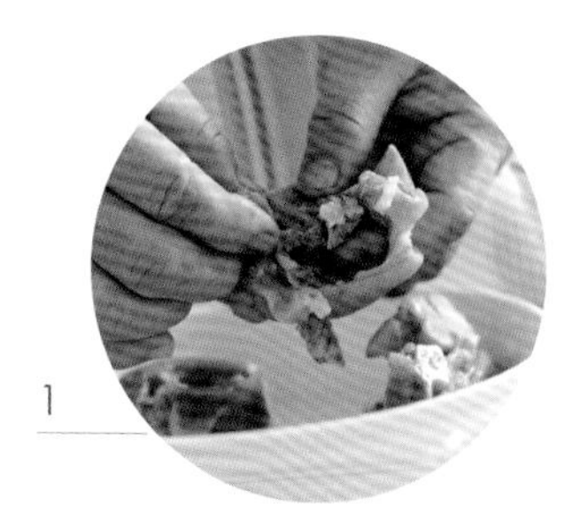
1

2

3

4

豬手凍

Ingredients (Serves 6)

1 frozen pork knuckle (about 900 g)
4 chicken feet
1 tbsp rock sugar
1 tbsp Shaoxing wine
2 tsp salt

Chinese marinade

2 slices ginger
2 star aniseed
1 tsp Sichuan peppercorns
1 small slice cinnamon bark
1 nutmeg (crushed)
2 bay leaves

Dipping sauce (mixed well)

2 tbsp white vinegar
1 tsp finely chopped garlic
red chilli dices

Preparation

1. Defrost the pork knuckle in the lower chamber of the refrigerator. Chop into pieces. Cook with the chicken feet for about 10 minutes. Rinse in cold water. Remove.
2. Put the Chinese marinade in a soup filter bag. Set aside.

Method

1. Put the Chinese marinade in 6 cups of water and bring to the boil. Add the pork knuckle, chicken feet, rock sugar and Shaoxing wine. Cook for about 2 hours until the meat is tender. Put in the salt and cook for a while. Remove. Reserve 1.5 cups of the cooking sauce. Skim oil from the sauce.
2. When the pork knuckle cools down, remove the bones. Tear off the meat. Cut the skin into flakes with a pair of scissors. Put in a glass container. (pictures 1-3)
3. Filter the cooking sauce. Pour into the container. Refrigerate for about 4 hours until set. (picture 4)
4. Remove the pork knuckle jelly. Cut into pieces. Serve with the dipping sauce.

素燒冰豆腐

Frozen Tofu and Vegetable Stew

材料（6人份量）

冰豆腐2塊（做法見預先準備部份）
栗子肉15粒
銀杏20粒
乾冬菇6朵
雲耳1湯匙
薑4片

調味料

老抽1湯匙
生抽1湯匙
糖1茶匙
麻油2茶匙
水2杯

預先準備

1. 冬菇及雲耳用水浸軟，去蒂，洗淨。
2. 燒滾清水，放入栗子（水浸過栗子）煮至大滾，盛起，過冷河，去外衣。
3. 板豆腐洗淨，瀝乾水分，待 15 分鐘。豆腐上下重疊放於食物袋（以重量壓出水分），紮緊，放入冷藏格至硬身，可保存 1 個月。（圖 1-3）

做法

1. 冰豆腐解凍，用水沖洗或浸軟，壓乾水分，切塊，再壓乾水分。（圖 4-6）
2. 燒熱鑊，下油 2 湯匙，下薑片及栗子炒香，加入冬菇及雲耳炒勻，傾入調味料煮滾，轉慢火煮 20 分鐘，最後加入冰豆腐及銀杏，用慢火煮 10 分鐘即可。

零失敗技巧

Successful cooking skills

冰豆腐的口感如何？
冰豆腐的製法簡易，質感綿軟，吸收醬汁後，非常美味！

How does the frozen tofu taste?
Having a soft texture, frozen tofu is very tasty after absorbing sauce. It is also easy to make.

為何選用板豆腐製冰豆腐？
板豆腐質感較實，重疊後不容易散爛。

Why use firm tofu to make frozen tofu?
With a firm texture, it hardly breaks when overlapping.

哪裏購買冰豆腐？
沒時間自製冰豆腐，可向街市售賣豆腐的檔子預早兩天預訂。

Where to buy frozen tofu?
If you have no time to make it, you can reserve it two days in advance from the tofu stalls in the market.

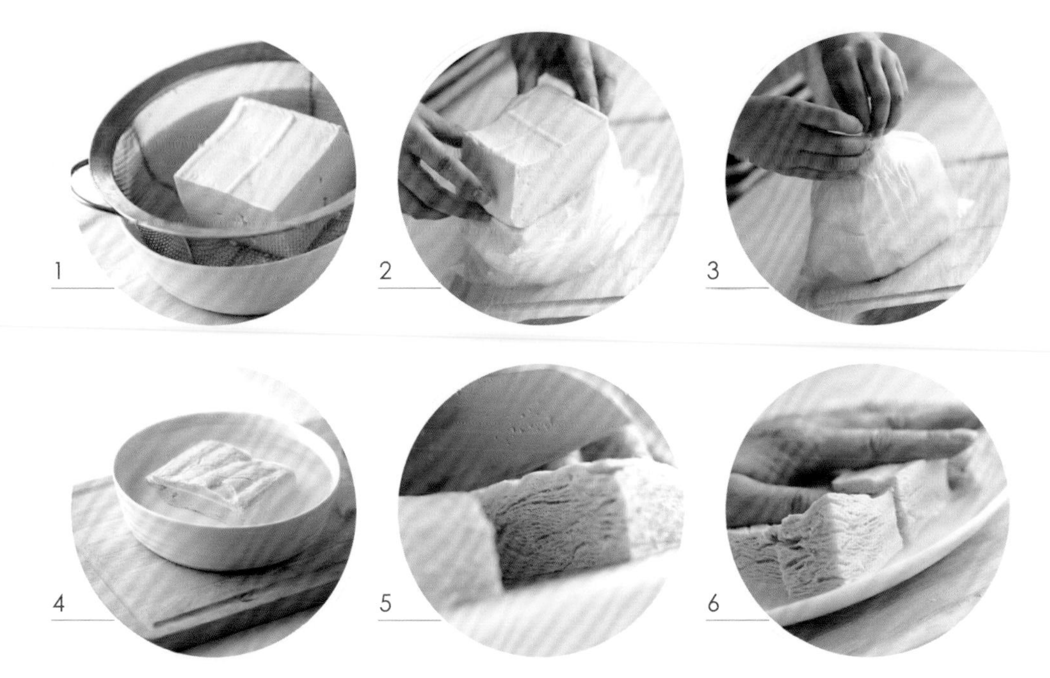

Ingredients (Serves 6)

2 pieces frozen tofu (refer to the preparation)
15 shelled chestnuts
20 ginkgoes
6 dried black mushrooms
1 tbsp dried cloud ear fungus
4 slices ginger

Seasoning

1 tbsp dark soy sauce
1 tbsp light soy sauce
1 tsp sugar
2 tsp sesame oil
2 cups water

Preparation

1. Soak the dried black mushrooms and cloud ear fungus in water until soft. Remove the stalks. Rinse.
2. Add the chestnuts in boiling water (water covering all the chestnuts). Cook until it heavily boils. Dish up and rinse with the cold water. Remove the skin.
3. Rinse firm tofu. Drain and set aside for 15 minutes. Overlap the beancurds and place them in a food bag (press water out). Tie the bag tightly. Put it in a freezer until firm. It can be preserved for 1 month. (pictures 1-3)

Method

1. Thaw the frozen tofu. Rinse or soak until soft. Squeeze dry. Cut into pieces. Squeeze dry again. (pictures 4-6)
2. Heat a wok. Put in 2 tbsp of oil. Stir-fry the ginger and chestnuts until fragrant. Add the mushrooms and cloud ear fungus. Stir-fry and mix well. Put in the seasoning and bring to the boil. Turn to low heat and cook for 20 minutes. Add the frozen tofu and ginkgoes at last. Cook for 10 minutes and serve.

士多啤梨生炒骨

Sweet and Sour Pork Ribs in Strawberry Tamarind Sauce

材料（6 人份量）

腩排 6 兩（斬塊）
士多啤梨 6 粒
青、黃甜椒各 1/4 個
羅望子半塊
乾粟粉半杯
油 1 湯匙

醃料
生抽 2 茶匙
紹酒 2 茶匙
胡椒粉少許

調味料
茄汁 2 湯匙
片糖 1.5 湯匙
鹽半茶匙

獻汁（拌勻）
粟粉 1 茶匙
水 3 湯匙

預先準備
1. 腩排洗淨，下醃料拌勻醃 1 小時。
2. 羅望子用滾水 1 量杯浸軟，過濾，隔渣備用。
3. 士多啤梨去蒂，洗淨，開邊；青、黃甜椒去籽，洗淨，切塊，用油炒熟備用。

做法
1. 腩排均勻地蘸上乾粟粉，放入滾油用慢火炸熟，再轉大火燒滾油炸至金黃色，盛起，隔去油分。
2. 羅望子醬及調味料用慢火煮滾，煮至片糖溶化，埋獻，下油拌勻，最後加入腩排、甜椒及士多啤梨輕拌即成。

零失敗技巧
Successful cooking skills

與常吃的生炒骨，味道有何分別？
此醬汁帶天然的果酸香味，而且酸甜美味，伴飯吃，美味之至！
How is this recipe different from regular sweet and sour pork?
This sauce is very fruity and tart. It is sweet and sour but with more depth and a complex palate. It is absolutely delicious.

羅望子哪裏有售？
羅望子於印尼雜貨店有售。
Where to buy dried tamarind?
Dried tamarind is available in the Indonesian store.

羅望子醬必須用慢火煮嗎？
耐心地用慢火輕輕煮勻，以免醬料焦燶，而且需不時攪拌。
Must the tamarind paste be cooked over low heat?
Yes, it must. You should cook over low heat patiently. Stir occasionally to prevent it from burning.

Ingredients (Serves 6)

225 g pork belly ribs (cut into cubes)
6 strawberries
1/4 green bell pepper
1/4 yellow bell pepper
1/2 slab dried tamarind seeds
1/2 cup cornflour
1 tbsp oil

Marinade

2 tsp light soy sauce
2 tsp Shaoxing wine
ground white pepper

Seasoning

2 tbsp ketchup
1.5 tbsp raw cane sugar slab (crushed)
1/2 tsp salt

Thickening glaze (mixed well)

1 tsp cornflour
3 tbsp water

Preparation

1. Rinse the pork ribs. Add marinade and mix well. Leave them for 1 hour.
2. Soak the tamarind seeds in 1 cup of boiling water until soft. Strain. Set aside.
3. Remove the stems of the strawberries. Rinse and cut into halves. Set aside. Seed the bell peppers. Rinse and cut into pieces. Stir fry the bell pepper in a little oil until done. Set aside.

Method

1. Coat the ribs evenly and lightly with cornflour. Deep fry in hot oil over low heat until done. Turn to high heat and fry until golden. Drain.
2. Boil the tamarind paste from step 2 and seasoning over low heat. Cook until the sugar dissolves. Stir in the thickening glaze. Stir in 1 tbsp of cooking oil. Put the ribs, bell pepper and strawberries back in. Toss well and serve.

越式燴羊膝

Vietnamese Style Lamb Shank Stew

材料（8 人份量）

羊膝 2 隻
洋葱 1 個
番茄 2 個
薑茸及蒜茸各 1 湯匙
紅辣椒 1 隻
香茅 1 枝
八角 2 粒
羊湯粒 1 粒
糖 2 茶匙
麵粉 1 湯匙

醃料

鹽 2 茶匙
胡椒粉少許

調味料

酸柑汁（泰國青檸）1 湯匙
魚露及鹽各適量

預先準備

1. 羊膝洗淨，抹乾水分，用醃料抹勻略醃。
2. 紅辣椒切幼粒；香茅、洋葱及番茄切碎。

做法

1. 燒熱少許油，羊膝均勻地抹上麵粉，下油鑊內略煎，加入蒜茸、薑茸、洋葱及紅辣椒炒透，加入糖略煎至帶焦香味。
2. 下香茅、八角、番茄、羊湯粒及水 3 杯，燜煮約 1.5 小時，至羊膝酥脸，最後加入調味料再煮至汁液略收乾即成。

零失敗技巧
Successful cooking skills

羊湯粒是甚麼？必須加入嗎？

羊湯粒是將羊肉熬湯後之精華，羊味濃郁可口，於超市有售。

What is a lamb stock cube? Must it be added into the dish?

It is made by cooking and condensing lamb soup into essence, which has an intense lamb flavour and tastes delectable. It is available at supermarkets.

爆炒洋葱等配料時，為何加入糖略拌？

洋葱與糖混合，會產生一種奇妙的效果，帶一份獨特的焦香味。

Why add sugar when stir-frying onion and other ingredients?

Onion combined with sugar creates a special effect which brings off a unique fragrant smell of burnt.

此菜的味道如何？

羊膝充滿香茅的清香、辣椒的微辣、番茄的甜香及青檸的酸甜，羊味濃！

What is the flavour of the dish?

The lamb shank gives a light aroma of lemongrass, a little spicy of chilli, a luscious taste of tomatoes, and a sour and sweet flavour of lime. It has a strong lamb flavour!

Ingredients (Serves 8)

2 lamb shanks
1 onion
2 tomatoes
1 tbsp finely chopped ginger
1 tbsp finely chopped garlic
1 red chilli
1 stalk lemongrass
2 star aniseed
1 lamb stock cube
2 tsp sugar
1 tbsp flour

Marinade

2 tsp salt
ground white pepper

Seasoning

1 tbsp Thai lime juice
fish gravy
salt

Preparation

1. Rinse the lamb shanks. Wipe dry. Spread with the marinade and rest for a while.
2. Finely dice the red chilli. Chop up the lemongrass, onion and tomatoes.

Method

1. Heat up a little oil. Spread the flour evenly on the lamb shanks. Slightly fry the lamb shanks. Add the garlic, ginger, onion and red chilli. Stir-fry thoroughly. Put in the sugar. Slightly fry until it smells the fragrance of burnt.
2. Put in the lemongrass, star aniseed, tomatoes, lamb stock cube and 3 cups of water. Simmer for about 1.5 hours until the lamb shanks are tender. Finally add the seasoning and cook until the sauce is a little bit dry. Serve.

洋葱炒黑毛豬腩片

Stir-fried Kurobuta Pork Belly with Onion

材料（6 人份量）

黑毛豬腩片 6 兩
洋葱 1 個
蒜香豆豉辣椒醬 2 茶匙（做法見 p.30）
紹酒半湯匙

醃料

生抽 2 茶匙
胡椒粉少許
粟粉 1 茶匙

調味料

糖半茶匙
生抽 1 茶匙

預先準備

1. 黑毛豬腩片洗淨，下醃料拌勻。
2. 洋葱去外衣，洗淨，切絲。

做法

1. 燒熱鑊下油 2 湯匙，下洋葱炒香，盛起。
2. 原鑊下辣椒醬及黑毛豬腩片炒勻，灒酒，加入洋葱、調味料及熱水 3 湯匙，不斷翻炒至豬腩片全熟即可享用。

Ingredients (Serves 6)

225 g thinly sliced Kurobuta pork belly
1 onion
2 tbsp black bean and garlic chilli sauce (refer to p.30)
1/2 tbsp Shaoxing wine

Marinade

2 tsp light soy sauce
ground white pepper
1 tsp cornflour

Seasoning

1/2 tsp sugar
1 tsp light soy sauce

Preparation

1. Rinse the pork belly. Add marinade and mix well.
2. Peel the onion. Rinse well and shred it.

Method

1. Heat a wok and add 2 tbsp of oil. Stir fry onion until fragrant. Set aside.
2. In the same wok, stir fry chilli sauce with pork belly slices. Sizzle with wine. Add onion, seasoning and 3 tbsp of hot water. Stir quickly until the pork is done. Serve.

蒜香豆豉辣椒醬
Black Bean and Garlic Chilli Sauce

材料
指天椒 3 兩
蒜茸、豆豉各 2 湯匙
蝦米 2 湯匙
粟米油 1.5 杯

調味料
老抽 1 湯匙
鹽 1 茶匙
糖 1.5 茶匙

做法
1. 指天椒洗淨，去蒂，切碎；蝦米洗淨，切碎；豆豉用水沖洗，切碎（或舂爛）。
2. 下油半杯燒熱，下蝦米及指天椒炒香，加入蒜茸及豆豉茸不斷炒香，下調味料及餘下之粟米油，用小火煮滾，再煮片刻，待涼，入瓶可冷藏 6 個月。

Ingredients
113 g bird's eye chillies
2 tbsp grated garlic
2 tbsp fermented black beans
2 tbsp dried shrimps
1.5 cups corn oil

Seasoning
1 tbsp dark soy sauce
1 tsp salt
1.5 tsp sugar

Method
1. Rinse the bird's eye chillies. Remove the stems. Finely chop them. Rinse the dried shrimps. Chop them. Rinse the fermented black beans. Chop or crush them. (Or you may crush ingredients separately with a mortar and pestle.)
2. Heat a wok and pour in 1/2 cup of oil. Stir fry dried shrimps and bird's eye chillies until fragrant. Add garlic and fermented black bean. Stir continuously until fragrant. Add seasoning and the remaining corn oil. Bring to the boil over low heat. Cook briefly. Leave it to cool. Store in sterilized bottles and refrigerate for 6 months.

零失敗技巧
Successful cooking skills

黑毛豬腩片的味道如何？

肥瘦均勻，肉質富彈性、爽口，沒有一般豬肉的肉腥味。

What's the taste of Kurobuta pork belly?

Kurobuta pork is a Japanese breed of black pig. Its belly has even marbling, with a unique crunch and chewiness. It's also free from the gamey taste that is associated with some pork.

如何確保豬腩片全熟卻又不會炒得過久？

注入熱水 3 湯匙，令豬腩片不必炒太久卻容易熟透。

How do you make sure the pork belly is cooked properly yet without being overcooked?

Add 3 tbsp of hot water after you put in the seasoning. The water speeds up the cooking process so that you don't have to stir fry it for too long.

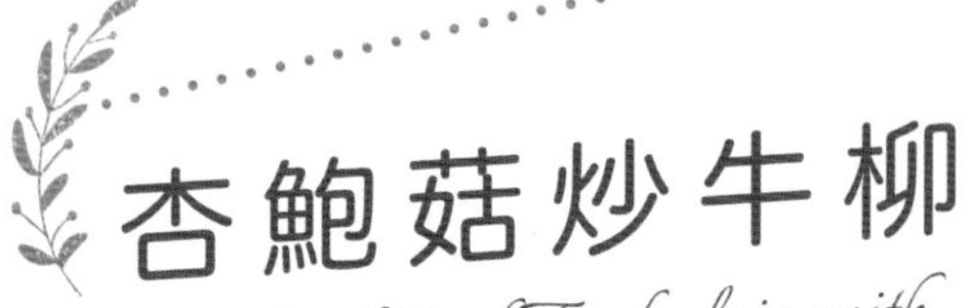

杏鮑菇炒牛柳

Stir-fried Beef Tenderloin with King Oyster Mushrooms

材料（6 人份量）

新鮮牛柳 4 兩
杏鮑菇 6 兩
蒜肉 2 粒
陳皮豆豉醬 1/3 湯匙
（做法見 p.34）
油 1 湯匙
紹酒半湯匙

醃料

黑胡椒碎 1 茶匙
生抽 1 茶匙
水 2 湯匙
粟粉 1 茶匙
油半湯匙（後下）

調味料

糖半茶匙
蠔油 1 茶匙
水 3 湯匙

預先準備

1. 牛柳切粗條，下醃料拌勻，醃 1 小時；下油再拌勻，備用。（圖 1-6）
2. 杏鮑菇洗淨，切薄片，飛水。

做法

1. 燒熱鑊下油 2 湯匙，放入牛柳拌散，炒至牛柳表面轉深色，盛起。
2. 原鑊下油 1 湯匙，下蒜肉及陳皮豆豉醬炒香，放入牛柳、杏鮑菇，灒酒，炒勻，最後下調味料炒片刻至牛柳全熟即成。

零失敗技巧

Successful cooking skills

如何切牛肉不粗韌？
順牛肉橫紋切，切斷肉筋，牛肉炒煮後不粗韌。

How do you slice the beef so that it won't be too chewy?
Slice across the grain (i.e. the blade is 90 degrees from the direction in which the muscle fibres run). You're essentially cutting short the muscle fibres and the beef will not be chewy and tough after being stir-fried.

用醃料醃牛柳後，為何最後才下油拌勻？
最後下油可緊封肉汁，而且炒煮時令牛柳容易散開，受熱均勻。

Why do you add oil to the beef at last after marinating the beef?
Adding oil helps seal in the beef juice. The sliced beef also opens up to be heated more evenly this way.

整個炒牛柳的過程需時多久？
約需時 7 至 8 分鐘。

How long does the stir-frying process last?
It takes about 7 to 8 minutes.

陳皮豆豉醬
Dried Tangerine Peel and Black Bean Sauce

材料
優質豆豉 2 兩
陳皮 1 個
蒜茸半湯匙
紅椒碎 1 茶匙

調味料
紹酒 1 湯匙
生抽 1 湯匙
老抽半湯匙
糖半湯匙
胡椒粉少許

做法
1. 陳皮用水浸軟，刮淨內瓤，切碎；豆豉用水沖洗，切碎，備用。
2. 燒熱鑊下油 3 湯匙，下豆豉炒至香，加入陳皮、蒜茸、紅椒碎及調味料炒至散發香味，待涼，入瓶可冷藏 2 個月。

Ingredients
75 g premium fermented black beans
1 whole dried tangerine peel
1/2 tbsp grated garlic
1 tsp chopped red chilli

Seasoning
1 tbsp Shaoxing wine
1 tbsp light soy sauce
1/2 tbsp dark soy sauce
1/2 tbsp sugar
ground white pepper

Method
1. Soak dried tangerine peel in water until soft. Scrape off the pith. Finely chop. Set aside. Rinse the fermented black beans in water. Finely chop them. Set aside.
2. Heat a wok and add 3 tbsp of oil. Stir fry black beans until fragrant. Add dried tangerine peel, grated garlic, red chilli and seasoning. Stir fry until fragrant. Leave it to cool. Store in sterilized bottles and refrigerate for 2 months.

1

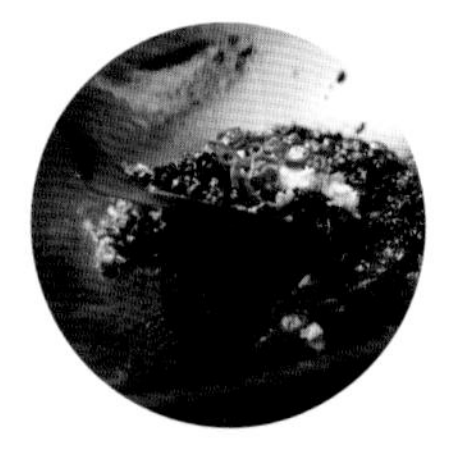
2

3

Ingredients (Serves 6)

150 g fresh beef tenderloin
225 g king oyster mushrooms
2 cloves garlic
1/3 tbsp dried tangerine peel and black bean sauce
1 tbsp oil
1/2 tbsp Shaoxing wine

Marinade

1 tsp crushed black pepper
1 tsp light soy sauce
2 tbsp water
1 tsp cornflour
1/2 tbsp oil (added at last)

Seasoning

1/2 tsp sugar
1 tsp oyster sauce
3 tbsp water

Preparation

1. Cut the beef tenderloin into thick strips. Add marinade and mix well. Leave it for 1 hour. Add oil and stir well. Set aside. (pictures 1-6)
2. Rinse the king oyster mushrooms. Slice thinly. Blanch in boiling water.

Method

1. Heat a wok and add 2 tbsp of oil. Put in the beef and scatter it. Stir fry until the beef is properly seared on all sides. Set aside.
2. In the same wok, heat 1 tbsp of oil. Stir fry garlic, dried tangerine peel and black bean sauce until fragrant. Put the beef and king oyster mushrooms. Sizzle with wine. Stir well. Lastly add seasoning and stir briefly until the beef is fully done. Serve.

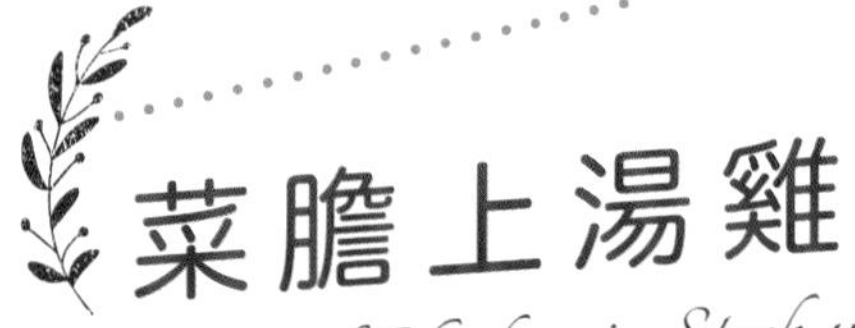

菜膽上湯雞

Steamed Chicken in Stock with Mustard Green

材料（12 人份量）

光雞 1 隻（約 2 斤至 2.5 斤）
芥菜膽 6 兩
清雞湯 2 杯（500 毫升）
葱 4 條
薑 3 片
紹酒 1 湯匙

調味料

蒸雞汁 1 1/3 杯（325 毫升）
鹽 1 茶匙
糖半茶匙

獻汁

蒸雞汁 2/3 杯（175 毫升）
鹽 3/4 茶匙
糖 1/3 茶匙
老抽半茶匙
麻油少許
生粉 2 茶匙

預先準備

光雞洗淨，放入熱水燙一會，盛起，沖洗乾淨，抹乾水分。

做法

1. 在雞腔內塗上紹酒，放入薑片及葱 2 條，餘下的葱切段，鋪在碟上。
2. 放上雞，注入清雞湯，隔水蒸半小時至雞熟透，預留 500 毫升蒸雞汁作調味及獻汁之用。
3. 芥菜膽修剪後，放入煮滾的調味料內，煮至軟身，隔去水分，上碟。
4. 雞待涼後，斬件，整齊排於碟上。燒滾獻汁，澆在雞件即成。

零失敗技巧
Successful cooking skills

蒸光雞前，為何先用熱水燙一會？
讓雞腔內的血水徹底沖洗乾淨，令賣相及味道更理想！
Why blanch the chicken in hot water before steaming?
This is to rinse the inside of the chicken thoroughly to remove blood, giving the chicken a better presentation and taste.

用清雞湯蒸雞，味道有很大分別嗎？
在蒸的過程中，令雞湯的雞味精華滲入雞肉，倍添美味！
Is there any difference in flavour by steaming the chicken in chicken stock?
It is tastier as the essence of chicken stock penetrates the meat in the steaming process.

為甚麼待雞涼透後才斬件？
若雞的溫度太熱，斬件時雞皮與雞肉容易分離，雞件散碎，賣相不美觀！
Why let the chicken cool down before chopping up?
The skin and meat will easily break up if chopping the chicken while it is hot. The broken pieces will make the dish look unpleasant!

Ingredients (Serves 12)

1 chicken (about 1.2 kg to 1.5 kg)
225 g trimmed mustard green
2 cups chicken stock (500 ml)
4 sprigs spring onion
3 slices ginger
1 tbsp Shaoxing wine

Seasoning

1 1/3 cups sauce from steaming chicken (325 ml)
1 tsp salt
1/2 tsp sugar

Thickening glaze

2/3 cup sauce from steaming chicken (175 ml)
3/4 tsp salt
1/3 tsp sugar
1/2 tsp dark soy sauce
sesame oil
2 tsp caltrop starch

Preparation

Rinse the chicken. Blanch in hot water for a while. Dish up. Rinse and wipe dry.

Method

1. Spread the Shaoxing wine on the chicken cavity. Put in the ginger and 2 sprigs of spring onion. Section the remaining spring onion. Lay on a plate.
2. Put the chicken on top. Pour in the chicken stock. Steam over water for 1/2 hour until fully done. Reserve 500 ml of sauce from steaming the chicken as the seasoning and thickening glaze.
3. Trim the mustard green. Put into the boiled seasoning and cook until soft. Dish up and drain. Arrange on a plate.
4. When the chicken cools down, chop into pieces. Arrange on the plate in order. Bring the thickening glaze to the boil. Sprinkle over the chicken. Serve.

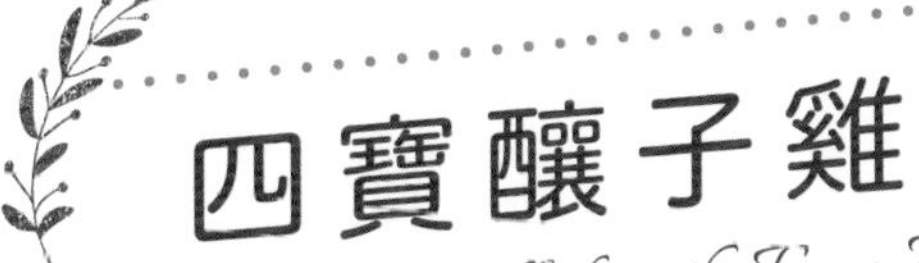

四寶釀子雞

Steamed Chicken Stuffed with Four Delicacies

材料（8 人份量）

鮮雞 1 隻（小，淨重約 1.5 斤）
糯米 4 兩
金華火腿 1 兩
瑤柱 3 粒
乾草菇半兩
紹酒 1 茶匙
薑汁酒 2 茶匙

醃料

老抽半茶匙
鹽 1 茶匙

調味料

蠔油 4 茶匙
糖及老抽各半茶匙
鹽 1/4 茶匙
麻油及胡椒粉各少許

預先準備

1. 糯米洗淨，用清水浸 4 小時，隔水蒸半小時。
2. 瑤柱沖淨，加入紹酒隔水蒸半小時，待涼，拆絲。
3. 乾草菇浸透、去蒂，切片；金華火腿抹淨，切片。

做法

1. 熱鑊下油，爆香乾草菇、瑤柱及金華火腿，加入糯米及調味料炒勻，待涼備用。
2. 用薑汁酒抹勻雞腔，釀入步驟（1）的餡料，用牙籤封口。
3. 雞皮用醃料抹勻，醃 1 小時。
4. 鮮雞放在碟上，隔水蒸 1 小時，隔去餘汁，待涼，剪開雞腹，取出餡料放在碟上；鮮雞斬件，伴餡料享用。

零失敗技巧
Successful cooking skills

乾草菇的味道，比乾冬菇優勝嗎？

乾草菇的菇味香濃，香味更勝乾冬菇，令雞腔及糯米等餡料，散發陣陣幽香的菇香氣味；但緊記徹底沖淨藏在草菇內的砂粒。

Are dried straw mushrooms better than dried black mushrooms in taste?

Dried straw mushrooms have a rich mushroom flavour and they are more fragrant compared with the dried black mushrooms. They bring the chicken and filling a delicate aroma of mushrooms. Rinse the raw mushrooms thoroughly to remove any sand grains.

步驟繁多，如何縮減時間？

蒸瑤柱及蒸糯米可同時進行，在糯米上架碟蒸瑤柱，可節省時間及燃料。

How to shorten the time for the complicated steps?

Steam the dried scallops and glutinous rice together. Put a steam rack over the glutinous rice and then place the dried scallops on top. It can save time and fuel.

宴客時，有哪些步驟可預早處理，令炮製更順暢？

瑤柱、乾草菇及金華火腿可於前一天預早處理，冷藏備用，以減省時間。

What steps can be done early to make the cooking smooth?

To save time, dried scallops, dried straw mushrooms and Jinhua ham can be prepared one day in advance and kept in a fridge.

Ingredients (Serves 8)

1 small fresh chicken (net weight about 900 g)
150 g glutinous rice
38 g Jinhua ham
3 dried scallops
19 g dried straw mushrooms
1 tsp Shaoxing wine
2 tsp ginger juice wine

Marinade

1/2 tsp dark soy sauce
1 Isp salt

Seasoning

4 tsp oyster sauce
1/2 tsp sugar
1/2 tsp dark soy sauce
1/4 tsp salt
sesame oil
ground white pepper

Preparation

1. Rinse the glutinous rice. Soak in water for 4 hours. Steam for 1/2 hour.
2. Rinse the dried scallops. Pour in the Shaoxing wine. Steam for 1/2 hour. Tear into shreds when cool.
3. Soak the dried straw mushrooms. Remove the stems and slice. Wipe and slice the Jinhua ham.

Method

1. Add oil in a heated wok. Stir-fry the straw mushrooms, dried scallops and Jinhua ham until fragrant. Put in the glutinous rice and seasoning. Stir-fry again. Let it cool down. Set aside.
2. Spread the ginger juice wine on the chicken cavity. Stuff with the filling from step (1). Seal with a toothpick.
3. Spread the marinade on the chicken skin. Rest for 1 hour.
4. Place the chicken on a plate. Steam for 1 hour. Sieve the sauce. Let it cool down. Cut open the chicken cavity with a pair of scissors. Remove the filling and place on the plate. Chop the chicken into pieces. Serve with the filling.

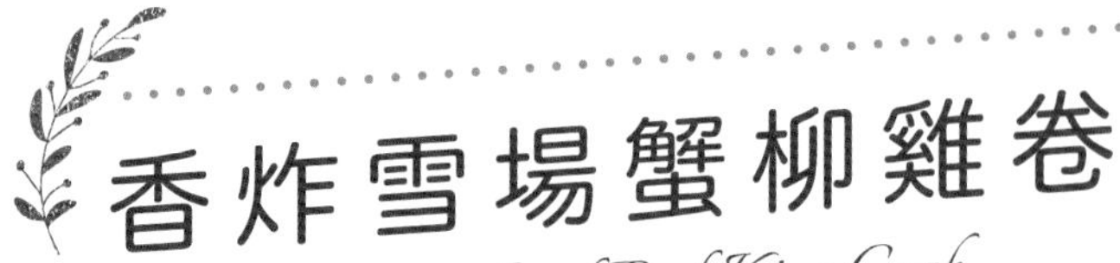

香炸雪場蟹柳雞卷

Deep-fried Red King Crab Meat Stick and Chicken Rolls

材料（12 人份量）

雞胸肉 10 兩
去皮馬蹄 3 粒
急凍雪場蟹柳 12 條
生粉適量

調味料

鹽 1/4 茶匙
生抽 2 茶匙
糖 1/8 茶匙
胡椒粉少許
生粉 1 湯匙

蘸汁

大紅浙醋適量

預先準備

1. 雞胸肉解凍、去皮，洗淨，吸乾水分；馬蹄拍碎，略剁備用。
2. 雞胸肉剁成茸，加入馬蹄碎及調味料拌至起膠，分成 12 等份。

做法

1. 將雞茸包裹一份蟹柳，保留部份蟹柳尖在餡外，捏成橢圓形，撲上生粉。
2. 放入熱油內炸至金黃色及熟透，蘸浙醋伴吃。

Ingredients (Serves 12)

375 g chicken breast
3 peeled water chestnuts
12 frozen red king crab meat sticks
caltrop starch

Seasoning

1/4 tsp salt
2 tsp light soy sauce
1/8 tsp sugar
ground white pepper
1 tbsp caltrop starch

Dipping sauce

Chinese red vinegar

Preparation

1. Defrost and skin the chicken breast. Rinse and wipe dry. Crush and roughly chop the water chestnuts. Set aside.
2. Chop the chicken breast into puree. Mix in the water chestnuts and seasoning. Stir into paste until sticky. Divide evenly into 12 portions.

Method

1. Wrap a crab meat stick in the chicken paste leaving the tip of the stick outside. Knead into an oblong shape. Coat with the caltrop starch.
2. Deep-fry in hot oil until golden and fully cooked. Serve with the Chinese red vinegar.

零失敗技巧
Successful cooking skills

雞胸肉的肉質會太粗韌嗎？
雞胸肉的膠質重，容易拌至起膠；用雞柳剁茸，適合老人家及小朋友食用，容易咀嚼。

Is the texture of chicken breast too tough?
Chicken breast is rich in collagen and can be easily stirred into paste. Minced chicken fillet is suitable for elders and kids as it is easier to chew.

用甚麼火候炸透？
先用大火，再用小火炸透，最後改大火迫出油分即可。

What degree of heat is used to?
Use high heat first. Then adjust to low heat to deep-fry thoroughly. Finally turn to high heat to press excessive oil out.

酸梅鴨

Braised Duck in Plum Sauce

材料（12 人份量）

冰鮮鴨或急凍鴨 1 隻
酸梅 10 粒（去核、搗碎）
麵豉醬 1 湯匙
蒜茸 1 湯匙
薑 6 片
紅辣椒半隻（切碎）
冰糖碎 1.5 湯匙
紹酒 1 湯匙
老抽 2 茶匙

調味料

生抽 1 茶匙
胡椒粉少許

預先準備

急凍鴨放於雪櫃的下層自然解凍，洗淨，去內臟，飛水，斬成 4 件，用老抽塗抹鴨身上。

做法

1. 燒熱少許油，放入鴨件略炸至金黃色（或在鴨身塗上少許油，放入焗爐烤 30 分鐘至外皮金黃）。
2. 燒熱少許油，加入蒜茸、薑片、紅椒碎、麵豉醬及酸梅茸炒透，下鴨件兜勻，灒紹酒，傾入水 1.5 杯、調味料及冰糖，燜約 45 分鐘至醬汁濃稠。
3. 取出鴨件，待涼，醬汁過濾，備用。
4. 鴨斬件，排於碟上，澆上醬汁即可。

Ingredients (Serves 12)

1 chilled or frozen duck
10 pickled plums (stoned; mashed)
1 tbsp ground bean sauce
1 tbsp finely chopped garlic
6 slices ginger
1/2 red chilli (chopped)
1.5 tbsp crushed rock sugar
1 tbsp Shaoxing wine
2 tsp dark soy sauce

Seasoning

1 tsp light soy sauce
ground white pepper

Preparation

Defrost the duck in the lower chamber of the refrigerator. Rinse and gut. Scald and chop into 4 pieces. Spread the dark soy sauce on the duck.

Method

1. Heat up a little oil. Deep-fry the duck until it is slightly golden (or spread a little oil on the duck and bake in an oven for 30 minutes until the skin is golden).
2. Heat up a little oil. Stir-fry the garlic, ginger, red chilli, ground bean sauce and pickled plum puree thoroughly. Put in the duck and stir-fry. Sprinkle with the wine. Pour in 1.5 cups of water, the seasoning and rock sugar. Simmer for about 45 minutes until the sauce reduces.
3. Remove the duck. Leave to cool down. Sieve the sauce and set aside.
4. Chop the duck into pieces. Arrange on a plate. Sprinkle with the sauce and serve.

零失敗技巧
Successful cooking skills

鴨件用油炸或烤焗，哪種方法較好？
將鴨件放入焗盤烤至金黃色，期間反轉一次再烤，做法較為方便，還可以將部份脂肪排出，少了一份油膩的感覺。

Which way to cook duck is better – deep-frying or baking?
Baking is easier. Put the duck on a baking tray and bake until golden. In the meantime, turn it over once and bake again. It can also help release part of the fat making the duck less greasy.

鴨肉咬起來帶嚼勁，有甚麼方法弄得更酥脍？
鴨件燜煮後，建議再蒸約 20 分鐘，令鴨肉酥軟之餘，也不會因煮得過久而焦燶。

Duck meat is chewy in texture. How to make it tenderer?
I suggest steaming the braised duck for about 20 minutes. It will tenderize the meat and prevent it from getting burnt when overcooked.

用酸梅 10 粒烹調，鴨肉的味道會太酸嗎？
當然不會！因為整隻鴨的份量大，而且混和了麵豉醬及冰糖等材料，酸甜適中，美味可口！

Will it be too sour by using 10 pickled plums?
Of course not! It is because the duck is large and mixed with ground bean sauce, rock sugar and other condiments. The sweet and sour taste is mild and wonderful!

鮮菌石斑球

Braised Grouper with Fresh Mushrooms

材料（6人份量）

石斑肉 8 兩
秀珍菇 3 兩
蟹味菇 2 兩
西芹 2 兩
甘筍 8 片
薑數片
蒜茸 1 茶匙
乾葱茸 1 茶匙
紹酒 2 茶匙

醃料

鹽 1/3 茶匙
胡椒粉少許
蛋白 1 湯匙
生粉半湯匙

調味料

鹽 1/3 茶匙
水適量

獻汁（拌勻）

水 4 湯匙
鹽 1/4 茶匙
蠔油半湯匙
糖 1/4 茶匙
麻油少許
胡椒粉少許
生粉 1 茶匙

預先準備

1. 石斑肉切件，加入醃料拌勻醃 10 分鐘。
2. 秀珍菇及蟹味菇切去根部，洗淨備用。
3. 西芹切成斜段。

做法

1. 秀珍菇及蟹味菇飛水，盛起，瀝乾水分。
2. 燒熱少許油，下西芹及調味料炒熟，盛起。
3. 石斑肉泡油，盛起。
4. 熱鑊下油，下薑片、蒜茸及乾葱茸爆香，加入全部材料，灒酒，注入獻汁煮滾，拌勻上碟。

零失敗技巧
Successful cooking skills

如何切魚球？

先預備利刀，將魚柳鋪平砧板上（魚皮朝下），用刀切每件約 3 厘米厚件，炒煮後魚皮收縮即成魚球。

How to make "curl-up" fish pieces?

Prepare a sharp knife and lay flat fish fillet on a chopping board (with the skin facing down). Slice fish into pieces of about 3cm thick. The fish skin would shrink after stir-fried, the fish pieces curled up and look like balls.

1

2

炒石斑肉有何竅門？

炒魚球必須輕力拌炒，否則魚肉容易鬆散。

What's the tip of stir-frying grouper flesh?

Grouper flesh must be stir-fried lightly or otherwise it will break into pieces.

如何去除鮮菇的異味？

烹調前緊記先飛水，可去掉鮮菇的異味。

How to remove unfavorable smell from fresh mushrooms?

Scald mushrooms before cooking can remove their unfavorable smell.

Ingredients (Serves 6)

300 g grouper flesh
113 g oyster mushrooms
75 g beech mushrooms
75 g celery
8 slices carrot
several slices ginger
1 tsp minced garlic
1 tsp minced shallot
2 tsp Shaoxing wine

Marinade

1/3 tsp salt
ground white pepper
1 tbsp egg white
1/2 tbsp caltrop starch

Seasoning

1/3 tsp salt
water

Thickening Sauce (mixed well)

4 tbsp water
1/4 tsp salt
1/2 tbsp oyster sauce
1/4 tsp sugar
sesame oil
ground white pepper
1 tsp caltrop starch

Preparation

1. Cut grouper flesh into pieces and marinate for 10 minutes.
2. Cut off roots from oyster mushrooms and beech mushrooms. Rinse and set aside.
3. Section celery at an angle.

Method

1. Scald oyster mushrooms and beech mushrooms. Drain.
2. Heat a little oil in a wok. Add celery and seasoning. Stir-fry until done and set aside.
3. Sizzle grouper flesh in oil for a while and drain.
4. Add oil into a hot wok. Stir-fry ginger slices, minced garlic and minced shallot until fragrant. Put in all ingredients. Pour in Shaoxing wine and the thickening sauce. Bring to the boil. Mix well and serve.

紅燜龍躉尾

Braised Giant Grouper Tail

材料（6人份量）
龍躉尾1斤4兩
乾冬菇4朵
半肥瘦豬肉3兩
薑4片
葱4條（切段）
紹酒1湯匙

醃料
胡椒粉少許
粟粉2茶匙

調味料
上等生抽1湯匙
鹽半茶匙
糖半茶匙

預先準備
1. 乾冬菇去蒂，用水浸軟，切絲；半肥瘦豬肉洗淨，切絲。
2. 龍躉尾洗淨，抹乾水分，下醃料拌勻，醃半小時。

做法
1. 燒熱鑊下油3湯匙，放入龍躉尾煎至兩面金黃，備用。
2. 原鑊下薑絲、豬肉絲及冬菇絲炒片刻，放入龍躉尾，灒酒，加入調味料及滾水1.5杯，燜約12分鐘至龍躉尾全熟及汁液濃稠，最後下葱段拌勻，上碟享用。

Ingredients (Serves 6)
750 g giant grouper tail
4 dried black mushrooms
113 g medium-fat pork
4 slices ginger
4 sprigs spring onion (sectioned)
1 tbsp Shaoxing wine

Marinade
ground white pepper
2 tsp cornflour

Seasoning
1 tbsp premium light soy sauce
1/2 tsp salt
1/2 tsp sugar

Preparation
1. Remove the stalks of the dried black mushrooms. Soak in water to soften. Cut into shreds. Rinse and shred the pork.
2. Rinse the giant grouper tail. Wipe dry. Mix with the marinade and rest for 1/2 hour.

Method
1. Heat up a wok. Add 3 tbsp of oil. Fry the giant grouper tail until both sides are golden. Set aside.
2. Put the ginger, pork and mushrooms in the same wok. Stir-fry for a while. Add the giant grouper tail. Sprinkle with the Shaoxing wine. Pour in the seasoning and 1.5 cups of boiling water. Simmer for about 12 minutes until the giant grouper tail is cooked through and the sauce thickens. Mix in the spring onion. Serve.

零失敗技巧
Successful cooking skills

甚麼時候的龍躉肉最好吃？

龍躉於年末秋冬季節最佳，肉質較肥美！

When does giant grouper taste best?

It tastes best in autumn and winter seasons when the flesh is the fattiest.

必須用上等生抽烹調嗎？

上等生抽的味道鮮美，配搭味鮮的龍躉尾，大大提升鮮味！

Is it necessary to cook with premium light soy sauce?

Premium light soy sauce is fresh and delicious. It greatly enhances the flavour of giant grouper which is already wonderful!

龍躉尾肉質太厚，容易燜熟嗎？

必須花點時間燜煮，或可用竹籤刺入魚肉，見無血水滲出即可。

The meat of the giant grouper tail is quite thick. Is it easy to be cooked through?

It takes some time to braise it. You may pierce the meat with a bamboo skewer to check its doneness. If no blood is oozing out, it is done.

乾冬菇去蒂後，才用水浸泡嗎？

可以！或浸泡後才剪掉冬菇蒂，較容易處理。

Can we soak the dried black mushrooms in water after removing the stalks?

Yes, you can. But it is easier to cut the stalks off after soaking.

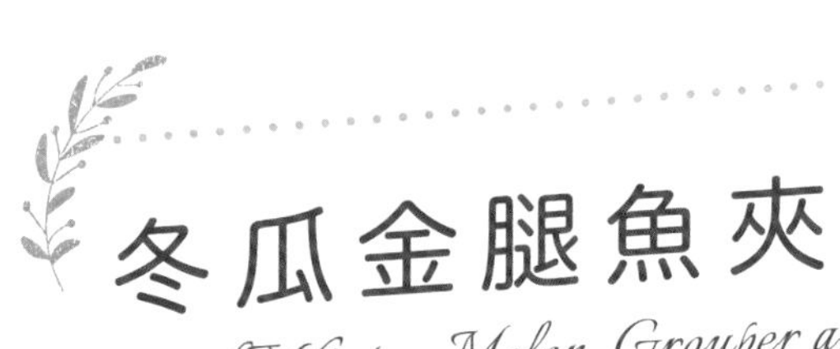

冬瓜金腿魚夾

Steamed Winter Melon, Grouper and Jinhua Ham Sandwiches

材料（12 人份量）
石斑肉 3 兩
冬瓜 10 兩
金華火腿 1 兩
清雞湯 125 毫升

醃料
鹽 1/8 茶匙
胡椒粉少許
蛋白 1 茶匙
生粉 1 茶匙

獻汁（拌勻）
清雞湯 100 毫升
糖 1/4 茶匙
麻油 1 茶匙
生粉 1 茶匙

預先準備
1. 冬瓜去皮，切成約 5cm×4cm×2cm 厚件（共 12 件），在瓜肉直切兩刀（不要切斷成雙飛狀）。（圖 1-4）
2. 燒滾水，下油 1 茶匙及冬瓜件煮 3 分鐘，取出，待涼。
3. 石斑肉切成 12 件，加入醃料拌勻。
4. 金華火腿切成 12 小片。

做法
1. 將斑肉及金華火腿慢慢釀入冬瓜內（或用小刀協助釀入），排在碟上，注入清雞湯蒸約 6 分鐘，隔去餘汁。（圖 5-8）
2. 煮滾獻汁，澆在冬瓜火腿魚夾上即可。

零失敗技巧
Successful cooking skills

石斑肉及金華火腿如何釀得美觀？
石斑肉及金腿的厚度必須切得較薄，高度比冬瓜件略短，釀入後更美觀。

How to stuff in grouper and Jinhua ham prettily?
The thickness of grouper slices and Jinhua ham slices should be thinner than that of wintermelon slices and their height should be a bit shorter than that of wintermelon.

應選購哪部份的冬瓜？
建議購買整個冬瓜的中間部份，瓜肉較軟身，容易釀入餡料。

Which part of wintermelon should be chosen?
Buy the middle part of wintermelon which is soft and easier for stuffing in fillings.

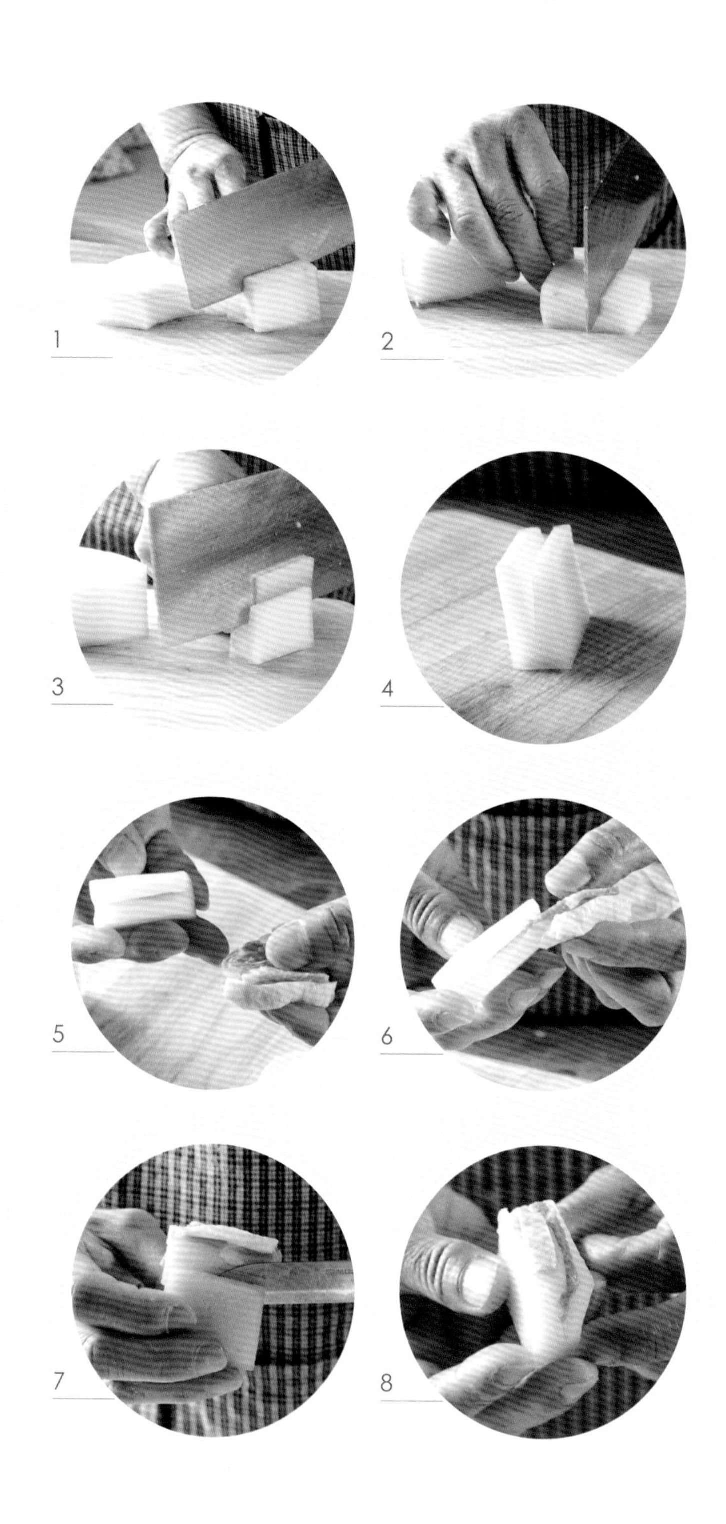
1
2
3
4
5
6
7
8

Ingredients (Serves 12)

113 g grouper flesh
375 g wintermelon
38 g Jinhua ham
125 ml chicken broth

Marinade

1/8 tsp salt
ground white pepper
1 tsp egg white
1 tsp caltrop starch

Thickening Sauce (mixed well)

100 ml chicken broth
1/4 tsp sugar
1 tsp sesame oil
1 tsp caltrop starch

Preparation

1. Skin wintermelon and cut into thick slices of sizes about 5cm x 4cm x 2cm (altogether 12 slices). Slit each slice at a side twice without cutting through. (pictures 1-4)
2. Bring water to the boil. Add 1 tsp of oil and cook the wintermelon for 3 minutes. Set aside to let cool.
3. Cut grouper flesh into 12 pieces and mix with the marinade.
4. Cut Jinhua ham into 12 small slices.

Method

1. Stuff grouper slices and Jinhua ham into the wintermelon slices slowly (or use a small knife for help). Arrange them into a plate. Pour in chicken broth and steam for about 6 minutes. Pour away the extract. (pictures 5-8)
2. Bring the thickening sauce to the boil and pour over the wintermelon. Serve.

三杯醬海鮮鍋

Seafood Hotpot in Three-cup Sauce

材料（6 人份量）

急凍帶子 4 兩
蝦 6 隻
鮮魷魚 1 隻
魚柳 4 兩
三色甜椒各半個
九層塔 2 棵
薑 8 片
乾葱 6 粒

三杯醬

麻油、紹酒、生抽各 3 湯匙
鎮江醋 2 湯匙
冰糖 1 湯匙（舂碎）

預先準備

1. 帶子解凍，洗淨；蝦剪去鬚及腳，挑腸，洗淨；鮮魷魚劏好，劃十字花，切塊；魚柳洗淨，切塊。
2. 三色甜椒去籽，洗淨，切塊。

做法

1. 燒熱瓦煲下油 1 湯匙，下薑片及乾葱炒香，加入鮮魷魚、魚塊及蝦拌勻。
2. 注入三杯醬煮滾，下三色甜椒及帶子煮至海鮮全熟，最後加入九層塔煮滾即成。

Ingredients (Serves 6)

150 g frozen scallops
6 prawns
1 fresh squid
150 g fish fillet
1/2 yellow bell pepper
1/2 green bell pepper
1/2 red bell pepper
2 sprigs Thai basil
8 slices ginger
6 cloves shallot

Three-cup sauce

3 tbsp sesame oil
3 tbsp Shaoxing wine
3 tbsp light soy sauce
2 tbsp Zhenjiang black vinegar
1 tbsp rock sugar (crushed)

Preparation

1. Thaw the scallops. Rinse well. Cut off the antennae and feet of the prawns. Devein and rinse well. Dress the squid and make light crisscross incision on the inside of the squid. Cut into pieces. Rinse the fish well and cut into chunks.
2. Seed the bell peppers. Rinse well and cut into pieces.

Method

1. Heat a clay pot. Add 1 tbsp of oil. Stir fry ginger slices and shallot until fragrant. Add squid, fish fillet and prawns. Stir well.
2. Pour in the three-cup sauce and bring to the boil. Add bell peppers and scallops. Cook until all seafood is done. Add Thai basil at last. Bring to the boil and serve.

零失敗技巧
Successful cooking skills

三杯醬料加入鎮江醋有何作用？

能增添海鮮的鮮味。

Why do you add Zhenjiang vinegar to the Three-cup sauce?

It brings out the seafood flavour very well.

用瓦煲烹調，有何注意之處？有何好處？

用瓦煲烹調，注意火力不宜太大，以免容易燒焦，而且用瓦煲煮出的菜式，香味特濃，非常惹味！

Is there anything that needs my attention when I cook with a clay pot? What are the advantages of cooking with clay pot?

Make sure you control the heat well. Clay pot conducts heat very quickly and the food may burn easily if cooked over high heat. Clay pot is usually related to rustic homely dish. It tends to make the food more flavourful with a rustic charm.

如何炒出爽口之帶子？

帶子最後才加入炒煮，炒至剛熟即可，以免久煮肉質過韌。

How do you make the scallops crunchy in texture?

Put them in at last and cook them until just done. Overcooking would make them tough and rubbery.

碧綠桂魚片

Stir-fried Mandarin Fish with Broccoli

材料（6 人份量）

桂花魚 1 尾（約 1 斤重）
西蘭花 8 兩
甘筍 8 片
薑汁酒 1 湯匙
蒜茸 1 茶匙

醃料

鹽半茶匙
胡椒粉少許
蛋白 2 湯匙
生粉 1 茶匙

調味料

水 3 湯匙
鹽半茶匙
糖 1/4 茶匙

獻汁

水 3 湯匙
蠔油半湯匙
糖 1/4 茶匙
麻油少許
生粉半茶匙

預先準備

1. 桂花魚起肉（魚販可代勞），切件，加醃料拌勻醃 10 分鐘。
2. 西蘭花切成小棵，飛水，盛起。

做法

1. 熱鑊下油，放入西蘭花及調味料拌炒，盛起。
2. 燒熱油，爆香蒜茸，放入桂花魚片炒熟，灒薑汁酒，下西蘭花、甘筍片及獻汁拌炒，滾後即可上碟。

Ingredients (Serves 6)

1 mandarin fish (about 600 g)
300 g broccoli
8 carrot slices
1 tbsp ginger wine
1 tsp minced garlic

Marinade

1/2 tsp salt
ground white pepper
2 tbsp egg white
1 tsp caltrop starch

Seasoning

3 tbsp water
1/2 tsp salt
1/4 tsp sugar

Thickening Sauce

3 tbsp water
1/2 tbsp oyster sauce
1/4 tsp sugar
sesame oil
1/2 tsp caltrop starch

Preparation

1. Bone mandarin fish (or ask the fish monger for help). Cut into slices and marinate for 10 minutes.
2. Cut broccoli into small floral pieces. Scald in boiling water and drain.

Method

1. Add oil into a hot wok. Put in the broccoli and seasoning. Stir-fry well and set aside.
2. Heat oil in wok. Stir-fry minced garlic until fragrant. Add the mandarin fish and stir-fry until done. Drizzle in ginger wine. Add broccoli, carrot slices and the thickening sauce. Mix well and cook until boiled. Serve.

零失敗技巧
Successful cooking skills

如何切魚片？1 斤重之桂花魚，可起出多少魚片？

預備利刀，魚柳鋪平砧板上（魚皮朝下），切每件約 1 厘米厚件。1 斤重桂花魚約可起出 6 兩淨肉。

How to slice fish? How many fish slices can be obtained from 600 g of mandarin fish?

Prepare a sharp knife and lay flat fish fillet on a chopping board (with the skin facing down). Slice fish into pieces of about 1cm thick. About 225 g of flesh from 600 g of mandarin fish.

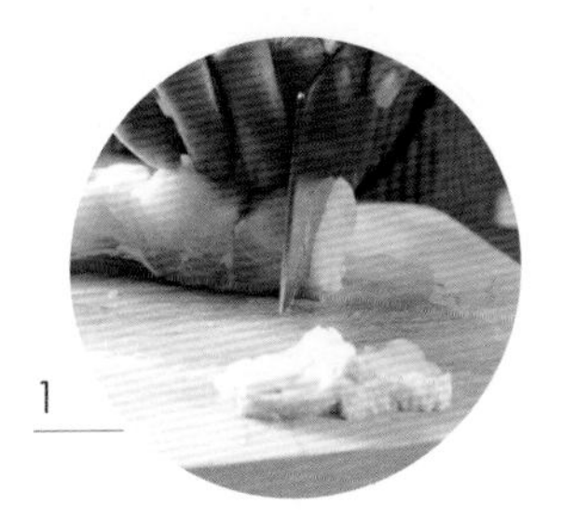

1

2

醃料內為何加入蛋白？

魚肉與蛋白拌勻，炒煮後魚肉仍保持嫩滑的口感。

Why egg white is used as one of the marinade sauce?

Fish mixed with egg white keeps smooth after stir-fried.

為何灒入薑汁酒？灒紹酒可以嗎？

薑汁酒可辟除魚腥味，適合炮製鮮魚菜式。

Why ginger wine is used? Can it be replaced with Shaoxing wine?

Ginger wine can swap away fishy taste and is best for fish dishes.

炒桂花魚片時，魚肉容易鬆散嗎？

建議在炒煮魚片時，勿太大力拌炒，以免魚片散碎。

Would the mandarin fish slices scatter when stir-fried?

It is suggested to stir-fry the fish slices lightly to avoid them scattered.

胡椒蝦

Fried Prawns with White Peppercorns

材料（6 人份量）

中蝦 12 兩
白胡椒粒 1.5 至 2 湯匙
紅辣椒 2 隻
淮鹽 1 茶匙
清水 3 湯匙

淮鹽料

幼鹽 1 湯匙
五香粉半茶匙

預先準備

1. 白鑊炒熱幼鹽，關火，加入五香粉炒勻，盛起備用。
2. 修剪蝦鬚及腳，挑去腸，洗淨，抹乾水分，用淮鹽醃 1 小時。
3. 白胡椒粒略舂碎；紅辣椒切圈。

做法

1. 熱鑊下油，加入中蝦略煎兩面，盛起。
2. 燒熱油，下白胡椒粒炒香，加入中蝦、紅辣椒及清水拌炒，加蓋焗煮片刻至汁液收乾，上碟，以羅勒裝飾享用。

Ingredients (Serves 6)

450 g medium-sized prawns
1.5 to 2 tbsp white peppercorns
2 red chillies
1 tsp spiced salt
3 tbsp water

Ingredients of spiced salt

1 tbsp fine salt
1/2 tsp five-spice powder

Preparation

1. Stir-fry table salt until hot in a wok without oil. Remove heat and mix in five-spice powder. Stir-fry well to become the spiced salt. Set aside.
2. Trim prawn tentacles and legs. Devein, rinse and wipe dry. Marinate with spiced salt for 1 hour.
3. Crush white peppercorns briefly. Cut red chillies into rings.

Method

1. Add oil into a hot wok. Fry both sides of medium prawns briefly. Set aside.
2. Heat oil in wok. Stir-fry white peppercorns until fragrant. Put in medium prawns, red chillies and water. Stir-fry well. Cover the lid and cook for a while until the sauce is dry. Garnish with basil and serve.

零失敗技巧
Successful cooking skills

如何令蝦殼香脆美味？

改用炸的方法，下中蝦略炸至蝦殼香脆，但耗油量比煎的方法略多。

How to make the prawn shells crispy and tasty?

Use deep-frying rather than frying can make prawn shells crispy but it uses more oil.

如何令白胡椒粒的香氣滲入蝦肉？

白胡椒粒炒香後，放入蝦加蓋焗煮，令胡椒香氣滲入蝦肉。

How to bring the smell of white peppercorns into the prawns?

Stir-fry white peppercorns until fragrant and then put in prawns. Cover the lid and cook for a while to bring the smell of white peppercorns into the prawns.

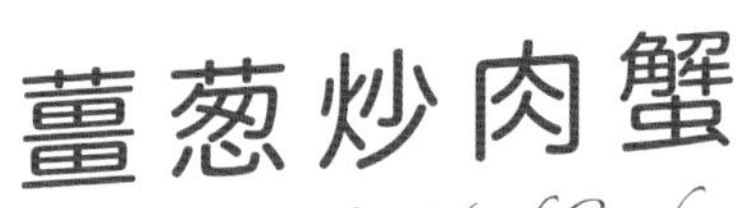

薑葱炒肉蟹

Stir-fried Male Mud Crab with Ginger and Spring Onion

材料（2 至 3 人份量）

肉蟹 1 隻（約 1 斤 4 兩）
老薑 1 塊
葱 6 條
紹酒半湯匙
粟粉 2 茶匙

調味料

鹽 1 茶匙
糖 1/4 茶匙

預先準備

1. 葱去鬚頭，洗淨，切段，分成葱白及青葱。
2. 老薑洗淨，用刀拍鬆。

做法

1. 肉蟹反轉背面，蟹腹向上，用刀在蟹身中間斬入（勿斬破蟹蓋），揭開蟹蓋，去掉蟹鰓及沙囊，洗淨，瀝乾水分，斬件，拍鬆蟹鉗，灑入粟粉拌勻。
2. 燒熱鑊下油 2 湯匙，下老薑及葱白爆香，加入蟹件炒片刻，灒酒炒勻，下調味料及滾水半杯，加蓋焗煮約 5 分鐘，待汁液收少，下青葱炒勻即成。

Ingredients (Serves 2-3)

1 male mud crab (about 750 g)
1 mature ginger
6 sprigs spring onion
1/2 tbsp Shaoxing wine
2 tsp cornflour

Seasoning

1 tsp salt
1/4 tsp sugar

Preparation

1. Remove the root of the spring onion. Rinse and cut into sections. Separate the white part from the green.
2. Rinse the mature ginger. Crush to loosen the flesh.

Method

1. Turn over the mud crab with the abdomen facing up. Chop into the middle part without cutting all the way through the shell. Lift the shell off. Remove the gills and stomach. Rinse and drain. Chop into pieces. Lightly crack the claws. Sprinkle with the cornflour and mix well.
2. Heat up a wok. Add 2 tbsp of oil. Stir-fry the ginger and white part of the spring onion until scented. Put in the crab and stir-fry for a moment. Sprinkle with the Shaoxing wine and stir-fry evenly. Pour in the seasoning and 1/2 cup of boiling water. Cook for about 5 minutes with a lid on. When the sauce reduces, put in the spring onion and stir-fry evenly to finish.

零失敗技巧
Successful cooking skills

如何挑選肉蟹？

眼睛及爪腳靈活運動；口吐泡沫；蟹身有重量；按其腹部有硬實飽脹的感覺；沒帶阿摩尼亞味的，表示為新鮮肉蟹。

How to select male mud crab?

Without the smell of ammonia, fresh mud crab has got active movement in the eyes, claws and legs with bubbles spilling from the mouth. It carries some weight and the abdomen is firm and plump to the touch.

為何選用老薑炒蟹？

老薑的薑味濃郁，毋須去皮，用刀拍鬆後，香味能徹底散發。

Why stir-fry the crab with mature ginger?

Mature ginger has a strong flavour. Crushing the ginger with the skin on helps fully spread its fragrance.

不懂斬蟹步驟，海鮮檔可代處理嗎？

相熟的店舖或可代處理；但想吃鮮活的炒蟹，還是自己用心學習處理吧！

We don't know how to chop a crab. Can the seafood stalls do this for us?

Your regular stalls may help, but learn it attentively if you want to eat fresh stir-fried crab!

蟹粉獅子頭

Crab Roe and Pork Balls with White Cabbage

材料（4 人份量）

肉蟹 1 隻（約 14 兩）
半肥瘦豬肉 4 兩
去皮馬蹄 2 個
小棠菜 4 兩
薑 2 片

調味料（1）

鹽及糖各 1/4 茶匙
生抽半湯匙
麻油及胡椒粉各少許
水 2 湯匙
生粉 1 湯匙

調味料（2）

清雞湯 125 毫升
清水 25 毫升
糖 1/4 茶匙
老抽 1 茶匙

生粉獻（拌勻）

水 1 湯匙
生粉 1 茶匙

預先準備

1. 肉蟹洗淨，取出蟹膏；蟹身隔水蒸約 13 分鐘，待凍，拆蟹肉約 2 至 3 兩。
2. 馬蹄用刀面拍碎，剁成幼粒。
3. 半肥瘦豬肉剁碎，與蟹肉 2 兩、蟹膏、馬蹄及調味料（1）拌勻，搓成 4 個獅子頭肉丸。
4. 小棠菜開邊，洗淨備用。

做法

1. 獅子頭肉丸用油半煎炸至表面硬身，盛起。
2. 燒熱砂鍋下油，爆香薑片，放入小棠菜，注入調味料（2）拌勻，排上獅子頭肉丸煮約 8 分鐘，埋獻，上碟品嘗。

零失敗技巧
Successful cooking skills

獅子頭的肉質會否硬實？

絕對不會，因選用半肥瘦豬肉及拌入水分混和，肉質綿軟及帶肉汁。

Would the pork balls taste firm in texture?

Absolutely not since medium-fat pork is used and mixed with water. Thus they taste soft and juicy.

市面有即拆蟹粉出售嗎？

有，於南貨店有售。

Are there instant crab roe sold in the market?

Yes. You can buy it at grocery stores.

Ingredients (Serves 4)

1 mud crab (about 525 g)
150 g medium-fat pork
2 peeled water chestnuts
150 g Shanghainese white cabbages
2 slices ginger

Seasoning (1)

1/4 tsp salt
1/4 tsp sugar
1/2 tbsp light soy sauce
sesame oil
ground white pepper
2 tbsp water
1 tbsp caltrop starch

Seasoning (2)

125 ml chicken broth
25 ml water
1/4 tsp sugar
1 tsp dark soy sauce

Caltrop starch sauce (mixed well)

1 tbsp water
1 tsp caltrop starch

Preparation

1. Rinse mud crab and set aside the crab roe. Steam the crab for about 13 minutes and set aside to let cool. Shell and bone to take about 75 g to 113 g of crabmeat.
2. Pat water chestnuts with a knife until mashed and chop into fine dices.
3. Chop medium-fat pork and mix with 75 g of crabmeat, crab roe, water chestnuts and seasoning (1). Knead the mixture into 4 round balls.
4. Cut white cabbages into two halves along the length. Rinse and set aside.

Method

1. Shallow-fry the pork balls with oil until the surfaces are hard in texture. Drain.
2. Heat oil in a clay pot. Stir-fry ginger slices until fragrant. Add white cabbages. Mix in seasoning (2). Arrange the pork balls into the pot and cook for about 8 minutes. Thicken with the caltrop starch solution. Serve.

芝士南瓜汁龍蝦球

Lobster Chunks with Cheese and Pumpkin Sauce

材料（4人份量）
龍蝦1隻（約1斤重）
中國南瓜 4 兩
芝士片 2 塊

醃料
鹽 1/4 茶匙
麻油及胡椒粉各少許
蛋白 1 湯匙
生粉 2 茶匙

調味料
鹽 1/8 茶匙
水 4 湯匙

預先準備

1. 處理龍蝦：於龍蝦尾部插入筷子，流出尿液。蝦身彎起，於頭及身之間的空隙切入，分開頭及蝦身。剪開蝦腹兩邊薄膜，小心地起出完整之龍蝦肉。(圖 1-8)
2. 蝦肉去腸，略沖，切件，用醃料拌勻醃 10 分鐘。(圖 9-10)
3. 南瓜去皮，切件，隔水蒸 10 分鐘，壓成南瓜茸，備用。

做法

1. 燒熱適量油，下龍蝦球泡油至熟，上碟。
2. 燒熱少許油，下南瓜茸、芝士片及調味料煮滾，至芝士片溶化，澆在龍蝦球上，趁熱品嘗。

零失敗技巧

Successful cooking skills

1 斤重之龍蝦，約可得多少龍蝦肉？

切去龍蝦頭後，約可取得 5 兩淨肉。

How much flesh can be obtained from 600 g of lobster?

About 188 g of flesh can be usually obtained from a lobster weighing 600 g after cutting the head.

在家不想用大量油泡油，怎辦？

龍蝦肉不泡油，烹調效果並不理想。若想省卻用油量，建議龍蝦肉分兩次泡油，可減省油量。

What can be done if I do not want to use large amount of oil for cooking lobster?

The effect is not good if lobster is not blanched in oil. To save the amount of oil used, you may cook lobster by two times.

可選用日本南瓜煮汁嗎？

日本南瓜的瓜肉較糯，配搭芝士烹調，口感膩滯，故建議選用中國南瓜。

Can Japanese pumpkin be used to cook the sauce?

Japanese pumpkin gives sticky mash and it tastes greasy if matched with cheese. Hence Chinese pumpkin is recommended.

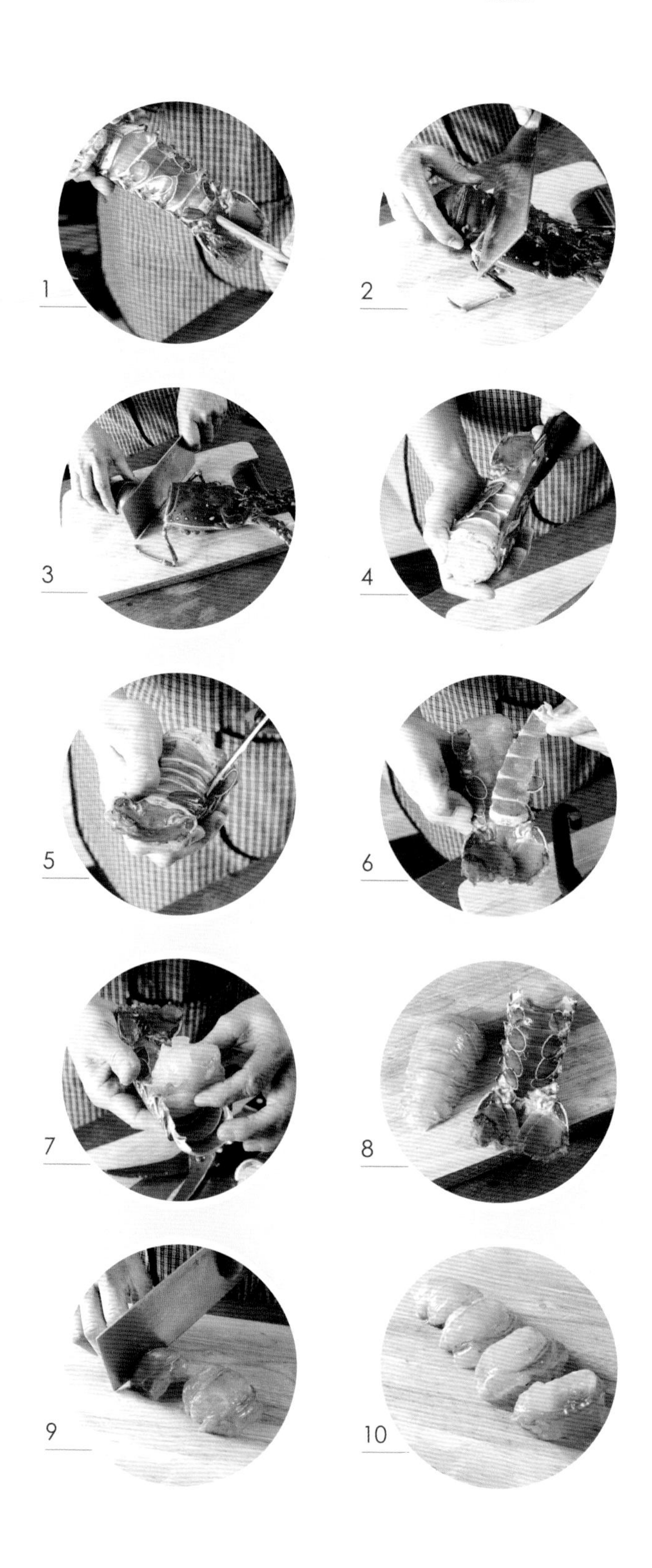
1
2
3
4
5
6
7
8
9
10

Ingredients (Serves 4)

1 lobster (about 600 g)
150 g Chinese pumpkin
2 slices cheese

Marinade

1/4 tsp salt
sesame oil
ground white pepper
1 tbsp egg white
2 tsp caltrop starch

Seasoning

1/8 tsp salt
4 tbsp water

Preparation

1. Preparing lobster: insert a chopstick into the lobster at its tail to let out urine from the lobster. Curl up the lobster body. Cut where the head joins the body with a knife. Cut open the thin membrane at the abdomen of lobster and remove the membrane. Take out the flesh in intact carefully. (pictures 1-8)
2. Remove entrails from lobster and rinse briefly. Cut into pieces and marinate for 10 minutes. (pictures 9-10)
3. Skin pumpkin and cut into pieces. Steam for 10 minutes and press into puree. Set aside.

Method

1. Heat oil in wok. Blanch the lobster in warm oil until done. Put on a plate.
2. Heat a little oil in a wok. Add pumpkin puree, cheese slices and seasoning. Bring to the boil until the cheese melted. Pour the sauce over the lobster and serve.

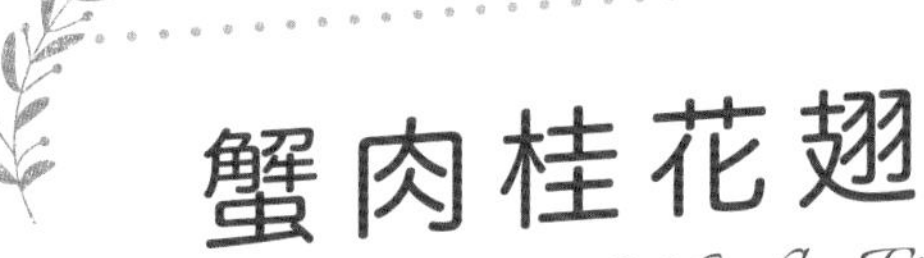

蟹肉桂花翅

Scrambled Eggs with Shark's Fin and Mung Bean Sprouts

材料（6 人份量）

急凍魚翅 5 兩
鮮蟹肉 2 兩
銀芽 2 兩
雞蛋 5 個
紹酒 1 茶匙
凍開水 1 湯匙

煨魚翅料

薑 2 片
葱 3 條
紹酒 1 茶匙
清水 750 毫升
油 1 湯匙

調味料（1）

鹽 1/4 茶匙

調味料（2）

鹽 3/4 茶匙
麻油及胡椒粉各少許
粟粉 3/4 湯匙
油 1 湯匙
紹酒半湯匙

預先準備

1. 魚翅解凍，拆絲。煮滾適量清水，放入魚翅煮腍，盛起，瀝乾水分。
2. 熱鑊下油，加入薑及葱爆香，灒酒，下清水、油 1 湯匙及魚翅煮 5 分鐘，盛起。

做法

1. 熱鑊下油，加入銀芽炒透，下調味料（1）炒勻，盛起，隔去水分。
2. 雞蛋及調味料（2）拂勻，加入蟹肉、銀芽及魚翅混合。
3. 燒熱油，傾入上述材料，灒紹酒，拌炒蛋液至將乾透，澆入凍開水，炒至蛋液乾身，上碟享用。

零失敗技巧
Successful cooking skills

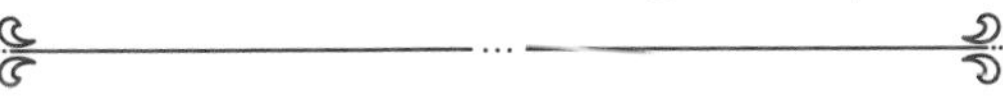

可用急凍蟹肉代替嗎？
絕對不可，因為鮮蟹肉是這道菜式之神髓，老饕可品嘗鮮甜之蟹肉香味。

Can fresh crabmeat be replaced with frozen crabmeat?
No, because the essence of this dish lies in fresh crabmeat that tastes fresh and sweet.

炒蛋液時，最後為何澆入凍開水？
灑入凍開水，令桂花翅不會太乾及質感太粗。

Why putting in cold drinking water at last when stir-frying the egg mixture?
Cold drinking water wets the shark's fin and egg mixture, it would not taste dry and rough in texture.

炒桂花翅的火候如何控制？
先用中火炒蛋液及蟹肉等材料，再調至小火炒至乾身，耐心處理必能成功。

How to control the heat when stir-frying the shark's fin and egg mixture?
Stir-fry the egg and crabmeat mixture over medium heat first and then turn to low heat and cook until dry. Patience can give you success.

Ingredients (Serves 6)

188 g frozen shark's fin
75 g fresh crabmeat
75 g mung bean sprouts
5 eggs
1 tsp Shaoxing wine
1 tbsp cold drinking water

Ingredients for cooking shark's fin

2 slices ginger
3 sprigs spring onion
1 tsp Shaoxing wine
750 ml water
1 tbsp oil

Seasoning (1)

1/4 tsp salt

Seasoning (2)

3/4 tsp salt
sesame oil
ground white pepper
3/4 tbsp cornflour
1 tbsp oil
1/2 tbsp Shaoxing wine

Preparation

1. Defrost shark's fin. Break it into shreds by hands. Bring water to the boil. Add shark's fin and cook until soft. Drain.
2. Add oil into a hot wok. Stir-fry ginger and spring onion until fragrant. Pour in Shaoxing wine lightly. Add water, 1 tbsp of oil and shark's fin. Cook for 5 minutes and set aside.

Method

1. Add oil into a hot wok. Stir-fry mung bean sprouts thoroughly. Add seasoning (1) and stir-fry well. Drain.
2. Whisk eggs and seasoning (2). Mix in crabmeat, mung bean sprouts and shark's fin.
3. Heat oil in wok. Pour in the above ingredients. Sizzle in Shaoxing wine. Stir-fry the egg mixture until breaks into bits and almost dry in texture. Add cold drinking water and stir-fry until dry in texture. Serve.

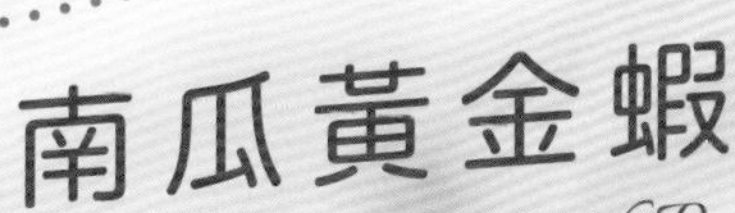

南瓜黃金蝦

Stir-fried Prawns with Pumpkin and Salted Egg Yolk

材料（6 人份量）

中蝦 300 克
南瓜 100 克（去皮、切片）
熟鹹蛋黃 3 個
生粉 2 湯匙
蒜茸 1 茶匙
牛油 1 湯匙
紹酒半茶匙

調味料

鹽 1/3 茶匙
胡椒粉少許

預先準備

1. 中蝦剪去蝦鬚及腳，於背部輕劃一刀，去掉蝦腸，抹乾備用。
2. 鹹蛋黃用叉子壓碎。

做法

1. 燒熱油 2 湯匙，中蝦及南瓜分別沾上生粉，煎至金黃色，盛起，瀝乾油分。
2. 燒熱牛油，下蒜茸及鹹蛋黃炒勻，待散發香味，加入調味料炒勻，輕推至起泡。
3. 放入中蝦及南瓜件輕拌沾滿鹹蛋黃茸，濳酒，快速兜勻，趁熱享用。

Ingredients (Serves 6)

300 g medium prawns
100 g pumpkin (skinned; sliced)
3 cooked salted egg yolks
2 tbsp caltrop starch
1 tsp finely chopped garlic
1 tbsp butter
1/2 tsp Shaoxing wine

Seasoning

1/3 tsp salt
ground white pepper

Preparation

1. Cut away the tentacles and legs of the medium prawns. Cut a slit in the back. Remove the veins. Wipe dry and set aside.
2. Mash the salted egg yolks with a fork.

Method

1. Heat up 2 tbsp of oil. Coat the medium prawns and pumpkin with caltrop starch separately. Fry until golden. Drain.
2. Heat up the butter. Stir-fry the garlic and mashed salted egg yolk evenly until fragrant. Add the seasoning and stir-fry until it bubbles.
3. Put in the medium prawns and pumpkin. Mix slightly until they are coated with the salted egg yolk. Sprinkle with the wine. Stir-fry swiftly. Serve.

零失敗技巧
Successful cooking skills

如何炒出香滑的鹹蛋黃茸？

必須用牛油起鑊，牛油的香氣與鹹蛋黃混和，香味倍增！

How to make stir-fried salted egg yolk puree fragrant and creamy?

Butter must be used. Heat up the butter and stir-fry with the salted egg yolk. The butter fragrance will greatly enhance the smell of the salted egg yolk.

可去掉中蝦殼，拌鹹蛋黃炒煮嗎？

不贊成！蝦殼能保留中蝦的鮮味，以免蝦肉直接觸及鑊邊，容易令蝦肉粗韌。

Can the medium prawns be shelled and then stir-fried with the salted egg yolk?

No! The shell can keep the fresh prawn flavour. It can prevent the meat from becoming tough by touching the wok directly.

南瓜容易散開來，怎辦？

煎南瓜時毋須太久，略煎即可，稍後再回鑊拌鹹蛋黃。

I am afraid the pumpkin will easily fall apart. How to do?

It is not necessary to fry the pumpkin for too long. Just give a slight frying. Return it to the wok later to cook with the salted egg yolk.

乾葱豉油焗中蝦

Fried Medium Prawns with Shallot and Light Soy Sauce

材料（6 人份量）

新鮮中蝦半斤
乾葱 4 粒
紹酒 1 湯匙
粟粉 1 茶匙

調味料

生抽半湯匙
老抽、黃砂糖各 1 茶匙
水 2 湯匙

預先準備

1. 中蝦剪去蝦鬚及腳，挑腸，洗淨，瀝乾水分，加入粟粉拌勻。
2. 乾葱去外衣，洗淨，切片。

做法

燒熱鑊下油 2 湯匙，放入中蝦煎至兩面金黃，下乾葱片爆香，灒酒，加入調味料煮片刻至汁液收乾，上碟食用。

Ingredients (Serves 6)

300 g fresh medium prawns
4 shallots
1 tbsp Shaoxing wine
1 tsp cornflour

Seasoning

1/2 tbsp light soy sauce
1 tsp dark soy sauce
1 tsp brown sugar
2 tbsp water

Preparation

1. Cut away the tentacles and legs of the medium prawns with a pair of scissors. Devein. Rinse and drain. Mix in the cornflour.
2. Peel, rinse and slice the shallots.

Method

Heat up a wok. Put in 2 tbsp of oil. Fry the medium prawns until both sides are golden. Stir-fry the shallots until fragrant. Sprinkle with the Shaoxing wine. Add the seasoning and cook for a while until the sauce dries. Serve.

零失敗技巧
Successful cooking skills

如何保存鮮蝦？

鮮蝦買回來後，建議用水完全浸泡，放入雪櫃冷藏，保存鮮蝦品質。

How to keep the medium prawns fresh?

Soak the prawns in water and put in a refrigerator to keep fresh.

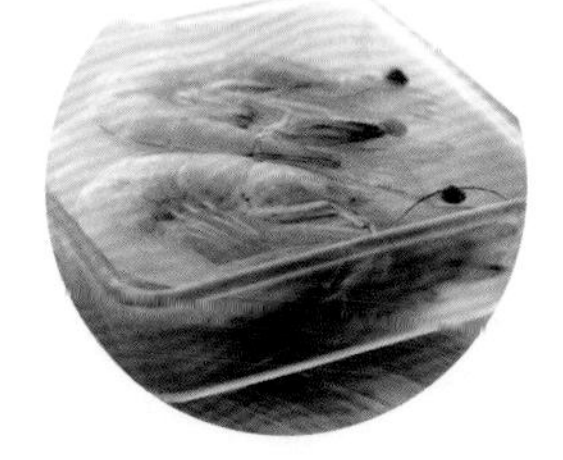

中蝦的價錢昂貴嗎？可用冰鮮蝦代替嗎？

中蝦售價不算太貴，每斤大約 80 元至 100 元。建議用鮮蝦烹調，你會吃到意想不到之鮮味！

Are medium prawns expensive? Can we use chilled ones?

They are not so expensive, costing around $80 to $100 per 600 g. It is better to use fresh medium prawns. Their taste is inconceivable!

家裏沒有黃砂糖，怎辦？

可用普通白糖代替，但黃砂糖帶一陣蔗糖原有之甜味，倍添鮮蝦原味。

There is no brown sugar at home. How to do?

You can use regular sugar but it lacks the raw sweetness of brown sugar to enhance the flavour of medium prawns.

鮑汁金銀帶子

Fried Scallops with Abalone Sauce and Dried Scallop Shreds

材料（8 人份量）
急凍大帶子 8 隻
乾瑤柱 2 粒
鮑魚汁 1 湯匙
紹酒 1 茶匙

醃料
鹽 1/4 茶匙
胡椒粉少許
生粉半茶匙
清雞湯適量

調味料
瑤柱汁 3 湯匙（若份量不足，可加水）
糖半茶匙
老抽 1 茶匙
麻油少許
生粉半茶匙

預先準備
1. 乾瑤柱沖淨，用暖水浸 2 小時，取出，與紹酒拌勻隔水蒸 45 分鐘（瑤柱汁留用），待涼後拆絲，下油炸脆。
2. 帶子解凍，沖淨抹乾，加入醃料（清雞湯除外）拌勻醃 15 分鐘，放入煮熱的清雞湯略浸（以浸過帶子為宜），盛起，吸乾水分。

做法
1. 燒熱油，放入帶子用慢火煎香兩面，上碟。
2. 煮滾鮑魚汁及調味料，澆在帶子上，灑入已炸脆之瑤柱絲即成。

零失敗技巧
Successful cooking skills

選用哪款帶子較合適？
建議選買日本北海道帶子，烹調後不會收縮。

Which kind of scallop should be used?

It is suggested to buy Japanese Hokkaido scallops that would not shrink after cooked.

帶子醃製後，為何放入熱湯略浸？
帶子略浸熱湯，能保持其形狀，而且減省煎帶子的時間，以免影響肉質變韌。

Why soak scallops in hot chicken broth after marinated?

Soaking scallops in hot chicken broth briefly can keep their shapes and also reduce their cooking time. Their texture not affected.

Ingredients (Serves 8)

8 frozen large scallops
2 dried scallops
1 tbsp abalone sauce
1 tsp Shaoxing wine

Marinade

1/4 tsp salt
ground white pepper
1/2 tsp caltrop starch
chicken broth

Seasoning

3 tbsp dried scallops extract (add water if the amount obtained is not enough)
1/2 tsp sugar
1 tsp dark soy sauce
sesame oil
1/2 tsp caltrop starch

Preparation

1. Rinse dried scallops and soak in warm water for 2 hours. Drain and mix with Shaoxing wine. Steam for 45 minutes and reserve the extract. Set steamed dried scallops aside to let cool. Tear into shreds. Deep-fry the shreds in oil until crispy.
2. Defrost scallops. Rinse and wipe dry. Mix with the marinade (except chicken broth) and set aside for 15 minutes. Soak in hot chicken broth briefly (cover the surface of the scallop). Drain and wipe dry.

Method

1. Heat oil in wok. Fry scallops over low heat until both sides are fragrant. Put on a plate.
2. Bring abalone sauce and seasoning to the boil. Pour over the scallops. Sprinkle over deep-fried dried scallop shreds and serve.

香草牛油焗翡翠螺

Baked Green Whelks with Herbs and Butter

材料（6 人份量）

翡翠螺 6 隻
煙肉 2 片（切碎）
洋葱碎 2 湯匙
蒜茸 2 茶匙
雜香草半茶匙
牛油 1 湯匙

調味料

鹽 1/3 茶匙
黑椒粉少許

預先準備

翡翠螺放入滾水內略灼，用小叉子將螺肉取出，去除內臟，切碎螺肉；螺殼留用。（圖 1-2）

做法

1. 燒熱牛油，加入洋葱碎及蒜茸炒香，下煙肉炒透，再放入螺肉拌炒。（圖 3）
2. 灑入雜香草及調味料炒勻，即成餡料。
3. 將餡料釀入翡翠螺殼內，放入已預熱的焗爐以 190℃焗約 25 分鐘，趁熱享用。（圖 4）

零失敗技巧
Successful cooking skills

翡翠螺肉很難取出嗎？
先將翡翠螺放於沸水灼片刻，螺肉收縮後，輕易取出。
Is it hard to take the green whelk meat out?
Scald the green whelk for a moment. The meat will shrink and you can easily take it out.

怎樣令餡料更惹味？
選用帶肥肉的煙肉作為餡料，與洋葱及蒜茸炒透後，香氣四散，別切掉肥肉！
How do you make the filling more flavourful?
Use fatty bacon in the filling. The fat adds an extra aroma after stir-fried with onion and garlic. Do not trim off the fat!

我喜歡芝士的香味，可灑點芝士烤焗嗎？
當然可以！芝士烤焗後令表面帶香脆口感！
I like the cheese aroma. Can I sprinkle some before baking?
Yes, of course! Baked cheese makes the surface crisp and smells fragrant.

Ingredients (Serves 6)

6 green whelks
2 slices bacon (chopped)
2 tbsp chopped onion
2 tsp finely chopped garlic
1/2 tsp mixed herbs
1 tbsp butter

Seasoning

1/3 tsp salt
ground black pepper

Preparation

Slightly blanch the green whelks. Take out the meat with a small fork. Remove the internal organs. Chop up the meat. Keep the shells. (pictures 1-2)

Method

1. Heat up the butter. Stir-fry the onion and garlic until fragrant. Add the bacon and stir-fry thoroughly. Put in the whelk meat and stir-fry. (picture 3)
2. Sprinkle with the mixed herbs and seasoning. Stir-fry evenly as filling.
3. Stuff the green whelk shells with the filling. Put in a preheated oven and bake at 190°C for about 25 minutes. Serve warm. (picture 4)

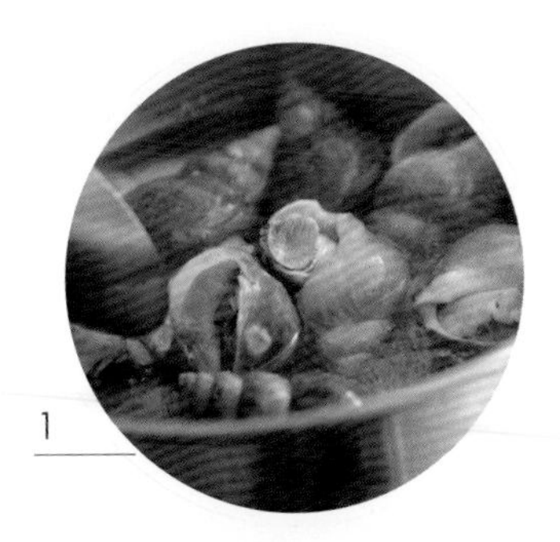
1

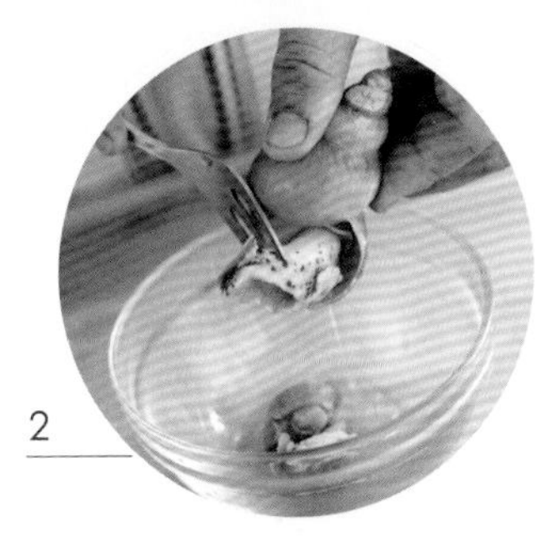
2

3

4

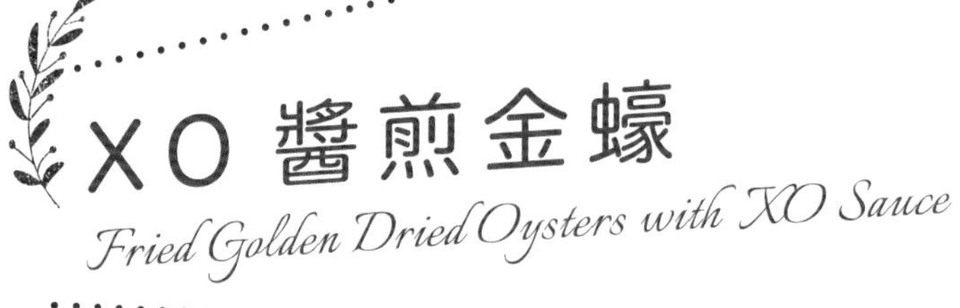

XO 醬煎金蠔

Fried Golden Dried Oysters with XO Sauce

材料（8 人份量）

金蠔 8 隻
XO 醬、薑汁酒各 1 湯匙

調味料

清水 4 湯匙
生抽、蜜糖各半茶匙

預先準備

金蠔洗淨，吸乾水分。

做法

1. 熱鑊下油，放入金蠔煎至兩面金黃色，盛起。
2. 燒熱油，下 XO 醬炒香，加入金蠔，灒薑汁酒，注入調味料用小火煮至汁液濃稠，上碟享用。

Ingredients (Serves 8)

8 golden dried oysters
1 tbsp XO sauce
1 tbsp ginger wine

Seasoning

4 tbsp water
1/2 tsp light soy sauce
1/2 tsp honey

Preparation

Rinse golden dried oysters and wipe dry.

Method

1. Add oil into a hot wok. Fry golden dried oysters until both sides are golden and set aside.
2. Heat oil in wok. Stir-fry XO sauce until fragrant. Add the golden dried oysters and pour in ginger wine at the side lightly. Put in the seasoning and cook over low heat until the sauce thickens. Serve.

零失敗技巧
Successful cooking skills

甚麼是金蠔？

金蠔是指曬至半乾濕之大蠔豉，海味店有售。

What is golden dried oyster?

Golden dried oyster is big oyster dried under the sun until medium dried. It can be bought from shops selling Chinese dried seafood stuff.

金蠔需要特別處理方法嗎？

不同店舖售賣的金蠔，軟硬程度有別，處理方法也不同：肉質較硬的金蠔，宜先蒸才烹調；較軟身的金蠔，洗淨後即可煎熟食用。

Do golden dried oysters require special preparation?

Golden dried oysters sold at different shops have different levels of hardness and thus require different handling ways. Harder golden dried oysters should be steamed before cooking while those soft in texture can be fried until done after rinsed.

甚麼是薑汁酒？可用紹酒代替嗎？

薑汁酒是薑汁與紹酒混和，比例為 1 比 1。薑汁酒可辟除金蠔的腥味。

What is ginger wine? Can it be replaced with Shaoxing wine?

Ginger wine is made by mixing ginger juice and Shaoxing wine in a 1:1 ratio. It can remove the fishy smell of golden dried oyster.

蛋白蒸海膽
Steamed Sea Urchin and Egg Whites

材料（6 人份量）
海膽 2 湯匙
蛋白 4 個
粟粉 1 茶匙
凍開水半杯

調味料
生抽、熟油各半湯匙

做法
1. 蛋白、粟粉及凍開水拌勻，用隔篩過濾，備用。
2. 將蛋白漿隔水慢火蒸 10 分鐘，鋪入海膽再蒸 1 分鐘，取出，吃時澆上調味料即成。

Ingredients (Serves 6)
2 tbsp sea urchin
4 egg whites
1 tsp cornflour
1/2 cup cold drinking water

Seasoning
1/2 tbsp light soy sauce
1/2 tbsp cooked oil

Method
1. Mix the egg whites, cornflour and cold drinking water together. Filter the egg mixture with a sieve. Set aside.
2. Steam the egg mixture over low heat for 10 minutes. Add the sea urchin and steam for 1 minute. Remove. Drizzle with the seasoning. Serve.

零失敗技巧
Successful cooking skills

海膽售價昂貴嗎？

一板海膽的售價由 70 至 120 元不等，想經濟實惠的話，可向壽司魚生店購買。

Is sea urchin expensive?

The price of sea urchin varies from $70 to $120 per plank. You can buy from sushi and sashimi shops, which are cheaper.

必須過濾蛋白漿嗎？

當然！隔去蛋白筋及粟粉顆粒，令蒸後的蛋白綿滑有緻，滑溜入口！

Need to filter the egg whites?

Of course! It helps remove the white tough substance and cornflour grains, giving a delicate texture to the steamed egg whites. It is so smooth!

宴請親友時，有甚麼需要預先準備嗎？

毋須預先準備，因此餸製法簡單，可即吃即煮。

I want to make this dish for a dinner party. What needs to be prepared in advance?

You don't need to prepare anything in advance. This recipe is quick and easy. You can serve it right away after making it.

蒸蛋白有何要訣？

蒸蛋白必須用慢火來蒸，蛋白嫩滑可口；海膽只需蒸熟即可，時間毋須太久。

What is your secret trick to velvety steamed custard?

Steam the custard over low heat to avoid overcooking. The sea urchin also cooks very quickly. Do not steam it for too long.

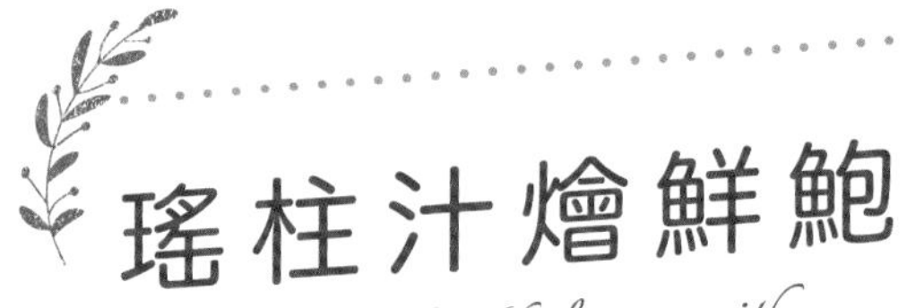

瑤柱汁燴鮮鮑

Braised Abalones with Dried Scallops Extract

材料（8 人份量）

大連鮮鮑魚 8 隻（約 14 兩）
乾瑤柱 1 兩
葱段 6 條
薑絲 1 湯匙
紹酒 1 湯匙

調味料

瑤柱汁 5 湯匙
鹽 1/4 茶匙
糖 1/4 茶匙
生粉 3/4 茶匙

預先準備

1. 乾瑤柱沖淨，用溫水 9 湯匙浸 2 小時，取出，與紹酒 1 湯匙拌勻，隔水蒸 45 分鐘，預留瑤柱汁 5 湯匙，瑤柱待涼，取部分瑤柱拆絲（約 2 湯匙）。
2. 鮮鮑魚連殼刷洗乾淨，用廚房剪刀起出鮑魚肉，去內臟，洗淨，抹乾水分。
3. 葱切絲備用。

做法

1. 鮑魚排於碟上，薑絲及葱絲放在鮑魚上，用大火蒸 6 分鐘，取出。
2. 煮滾少許油及調味料拌勻，下瑤柱絲拌勻，澆在鮑魚上品嘗。

零失敗技巧
Successful cooking skills

大連鮮鮑魚價錢昂貴嗎？

大連鮮鮑魚屬中價海鮮，一斤售大約港幣 120 元至 130 元，每斤之數量視乎鮑魚大小而定。

Are Dalian fresh abalones expensive?

They belongs to medium priced seafood with abut HK$120 to 130 for 600 g. The number of abalones for 600 g depends on the size of abalones.

選用大顆瑤柱或小顆的，對食味有影響嗎？

只要是優質的瑤柱，瑤柱碎也絕無問題。

What size of dried scallops should be used and would the taste of the dish be affected?

Use good-quality dried scallops and even fragments of dried scallops can do.

Ingredients (Serves 8)

8 Dalian fresh abalones (about 525 g)
38 g dried scallops
6 sprigs spring onion (sectioned)
1 tbsp shredded ginger
1 tbsp Shaoxing wine

Seasoning

5 tbsp dried scallops extract
1/4 tsp salt
1/4 tsp sugar
3/4 tsp caltrop starch

Preparation

1. Rinse dried scallops and soak in 9 tbsp of warm water for 2 hours. Drain and mix well with 1 tbsp of Shaoxing wine. Steam for 45 minutes. Reserve 5 tbsp of the dried scallops extract. Set the dried scallops aside to let cool. Tear some dried scallops into shreds (about 2 tbsp).
2. Wash and brush fresh abalones with their shells. Take up the abalones with a pair of kitchen scissors. Remove entrails. Rinse and wipe dry. Put the abalones on the shells.
3. Shred spring onion and set aside.

Method

1. Put abalones on a plate. Put ginger shreds and spring onion shreds over the abalones. Steam over high heat for 6 minutes.
2. Bring a little oil and seasoning to the boil. Mix well and add dried scallops shreds. Pour over the abalones and serve.

背煎豉汁蟶子皇

Fried Razor Clams in Black Bean Sauce

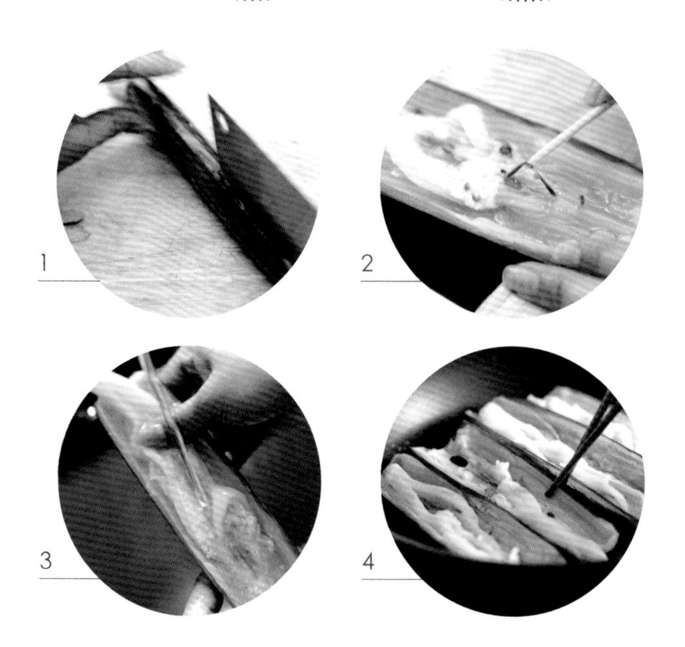

材料（8 人份量）

蟶子皇 8 隻
蒜茸 1 湯匙
豆豉茸 1 茶匙
紅椒絲少許

調味料（調勻）

生抽半湯匙
老抽 1 茶匙
糖 1/4 茶匙

預先準備

蟶子皇擦淨外殼，用刀在蟶子皇的中央位置直剘一刀，挑淨腸臟，洗淨，抹乾水分。（圖 1-3）

做法

1. 平底鑊下油 2 湯匙，排上蟶子皇（外殼向下），用中火煎至蟶子肉收縮，上碟。（圖 4）
2. 原鑊加入蒜茸及豆豉茸爆香，下紅椒絲及調味料煮滾，澆在蟶子皇上，趁熱食用。

Ingredients (Serves 8)

8 large razor clams
1 tbsp finely chopped garlic
1 tsp finely chopped fermented black beans
shredded red chilli

Seasoning (mixed well)

1/2 tbsp light soy sauce
1 tsp dark soy sauce
1/4 tsp sugar

Preparation

Rub the razor clam shells clean. Gut the razor clams by straightly cutting into the middle of the body. Remove the intestine with the skewer. Rinse and wipe them dry. (pictures 1-3)

Method

1. Put 2 tbsp of oil in a pan. Line the razor clams on the pan with the shell down. Fry over medium heat until the razor clam meat shrinks. Put on a plate. (picture 4)
2. Stir-fry the garlic and fermented black beans in the same pan until scented. Add the red chilli and seasoning. Bring to the boil. Drizzle on the razor clams. Serve warm.

零失敗技巧
Successful cooking skills

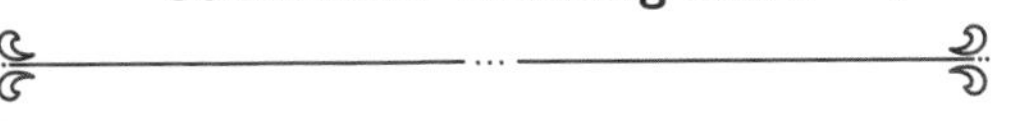

甚麼是背煎法？

背煎法是將貝類的外殼排於鑊底煎煮，避免蟶子肉直接觸及明火，令肉質嫩滑，味道更鮮更濃。

What is the method of frying shellfish in half shell?

Put the shellfish in a pan with the shell down and the meat up. It can prevent the meat from direct heating, making it more smooth and delicious.

如何確定蟶子肉熟透？

見蟶子肉呈收縮狀即表示剛熟，必須立即上碟；否則蟶子肉久煮，肉質變韌。

How do you make sure the razor clams are cooked through?

They are cooked as soon as they shrink. You should remove them from heat right away. Otherwise, they will be overcooked and turn rubbery.

蟶子皇經常有售嗎？

蟶子皇全年皆有供應，但以夏末秋初之時，肉質最肥美。

Are large razor clams always available?

They are sold year-round, and most plump in summer end and early fall.

蟶子的鮮甜會被蒜茸及豆豉掩蓋嗎？

不會！將醬汁輕輕澆上蟶子肉上品嘗，更能帶出蟶子的鮮美！

Do the garlic and fermented black bean overpower the razor clam in flavour?

No! The razor clam meat seasoned with a light sauce tastes particularly sweet and sensational!

帶子百花石榴粿

Steamed Scallops and Prawns Beggar's Purses

蛋白皮材料（8 人份量）
蛋白 8 個
生粉 2 茶匙
清水 2 湯匙

蛋白皮調味料
鹽 1/3 茶匙

石榴粿餡料
帶子 4 隻
蝦肉 3 兩（去腸）
去皮馬蹄 2 粒
西芹 1 兩

餡料調味料
鹽 1/3 茶匙
胡椒粉少許
麻油 1 茶匙

配料
韮菜花 8 條

獻汁
清雞湯 6 湯匙
麻油少許
生粉 1 茶匙

預先準備
1. 蛋白皮材料拌勻，拌入調味料。平底鑊燒熱油，下蛋白料煎成薄薄蛋白皮（共 8 張），待涼備用。
2. 帶子、蝦肉、馬蹄及西芹切粒。熱鑊下油，下餡料拌炒，加入調味料炒熟，分成 8 份，待涼。
3. 韮菜花放入滾水飛水至軟身，備用。

做法
1. 蛋白皮鋪平，放入適量餡料包摺成石榴粿，用韮菜花包紮，放於碟上，隔水蒸 5 分鐘，取出。（圖 1-5）
2. 煮滾獻汁，澆在石榴粿上即可享用。

Ingredients of egg white wrappers (Serves 8)

8 egg whites
2 tsp caltrop starch
2 tbsp water

Seasoning of egg white wrappers

1/3 tsp salt

Fillings of beggar's purses

4 scallops
113 g shelled prawns (deveined)
2 peeled water chestnuts
38 g celery

Seasoning of fillings

1/3 tsp salt
ground white pepper
1 tsp sesame oil

Condiment

8 flowering Chinese chives

Thickening sauce

6 tbsp chicken broth
sesame oil
1 tsp caltrop starch

Preparation

1. Mix the ingredients of egg white wrappers. Add seasoning and mix well. Heat oil in a frying pan. Fry 8 thin egg white wrappers and set aside to let cool.
2. Dice scallops, shelled prawns, water chestnuts and celery. Add oil into a hot wok. Put in the fillings and stir-fry well. Add seasoning and stir-fry until done. Divide into 8 portions and set aside to let cool.
3. Scald flowering Chinese chives in boiling water until soft and set aside.

Method

1. Lay flat the egg white wrappers and put in the fillings. Wrap the beggar's purses and fix the ends with flowering Chinese chives. Put on a plate and steam for 5 minutes. (pictures 1-5)
2. Bring the thickening sauce to the boil and pour over the beggar's purses. Serve.

零失敗技巧
Successful cooking skills

蛋白皮煎後，可放置一會才包裹餡料嗎？

建議蛋白皮煎後即包入餡料，否則蛋白皮皺摺，影響外觀。

Can egg white wrappers set aside for a while before wrapping?

It is recommended to wrap in fillings right after frying egg white wrappers, or they wrinkle and affect the appearance.

煎蛋白皮有何竅門？

1. 火候細；2. 油分少；3. 蛋白漿塗勻平底鑊。

What are the tips of frying egg white wrappers?

1. Fry over low heat; 2. use little oil only; 3. lay the egg white paste evenly in the frying pan.

蛋白皮料內為何加入生粉？

增加蛋白皮的韌度，捏摺時不容易弄破。

Why add caltrop starch into the egg white wrappers ingredients?

It increases the hardness of egg white wrappers, so they do not break easily during wrapping.

韭菜花容易折斷，怎辦？

韭菜花先飛水才包紮，只要小心處理絕無問題，同時可多預備韭菜花以作不時之需。

How to do with the easy-broken flowering Chinese chives?

Scald chives beforehand and handle carefully. Or you may also prepare more chives.

海膽菜粒炒飯

Stir-fried Rice with Sea Urchin and Vegetable

材料（4 人份量）

海膽 2 湯匙
菜莖 6 條
蝦乾 1 湯匙
雞蛋 1 個（拂勻）
薑茸 2 茶匙
白飯 2 碗

調味料（調勻）

幼海鹽 1 茶匙
水 1 湯匙

預先準備

1. 蝦乾用水浸 15 分鐘，盛起，切粒。
2. 菜莖洗淨，切粒，飛水，瀝乾水分。

做法

1. 燒熱鑊下油 2 湯匙，下薑茸及蝦乾爆香，盛起。
2. 原鑊下油 1 湯匙，傾入蛋漿及白飯，炒至飯粒鬆散，加入薑茸、蝦乾、菜莖及調味料炒勻，待飯粒熱透，最後加入海膽炒片刻即成。

Ingredients (Serves 4)

2 tbsp sea urchin
6 vegetable stems
1 tbsp dried prawns
1 egg (beaten)
2 tsp finely chopped ginger
2 bowls steamed rice

Seasoning (mixed well)

1 tsp fine sea salt
1 tbsp water

Preparation

1. Soak the dried prawns in water for 15 minutes. Dish up and dice.
2. Rinse and dice the vegetable stems. Blanch and drain.

Method

1. Heat up a wok. Put in 2 tbsp of oil. Stir-fry the ginger and dried prawns until scented. Remove.
2. Add 1 tbsp of oil in the same wok. Pour in the egg wash and rice. Stir-fry until the rice loosen. Add the ginger, dried prawns, vegetable stems and seasoning. Stir-fry and mix well. When the rice is warmed through, add the sea urchin and stir-fry for a moment. Serve.

零失敗技巧
Successful cooking skills

可以用哪種菜莖？
菜心、芥蘭、西蘭花等蔬菜皆可。

Use stems of which vegetables?

Chinese flowering cabbage, kale or broccoli will do.

必須用前一天煲煮的白飯嗎？
建議使用前一天的白飯，飯粒清爽、水分少，炒起來飯粒分明；或可於當天煲煮白飯，但必須減少水量。

Need to use the rice cooked yesterday?

It is recommended to use the leftover rice as they are dry and do not stick together during the frying process. You may also cook the rice before frying, but reduce the water used.

宴請賓客享用，如何做到色香味俱全？
蛋漿及白飯先炒透，飯粒裹上金黃的色澤；待飯粒炒至熱透香口，最後拌入海膽，令飯粒滲滿蝦乾及海膽的鮮香味，令賓客嘖嘖讚好！

I want to make this dish for dinner party. How can I make it look, smell and taste great?

Stir-fry the whisked eggs and day-old rice first to coat each grain evenly in the golden egg. When the rice is heated through and lightly browned, stir in the sea urchin at last to imbue each rice grain with the umami of dried shrimps and sea urchin. That's all it takes to wow your guests.

黃金豆腐醬伴脆米粉

Crispy Rice Vermicelli with Salted Egg Yolk and Tofu Paste

材料（4 人份量）

豆腐 1 塊
蝦仁 4 兩
鹹蛋 2 個
米粉 1 份

調味料

糖 1/3 茶匙
鹽半茶匙
粟粉 1 茶匙
麻油 2 茶匙
水 3 湯匙

預先準備

1. 鹹蛋（連殼）洗淨，放於清水加熱至滾，烚約 6 分鐘至熟；鹹蛋黃壓碎。（圖 1）
2. 蝦仁洗淨，飛水；豆腐用叉子壓碎。（圖 2）

做法

1. 米粉放入滾油內，炸片刻至鬆脆，隔油，上碟。
2. 燒熱鑊，下油 2 湯匙，下鹹蛋黃炒香及起泡，下豆腐及蝦仁煮片刻，加入調味料煮滾拌勻，盛起，澆在炸米粉上，拌勻食用。（圖 3-6）

零失敗技巧
Successful cooking skills

炒鹹蛋黃有何成功要訣？
耐性最重要！鹹蛋黃必須壓碎，用中小火慢炒至起泡。

How to make stir-fried egg yolks successfully?

Patience is the key! The salted egg yolks must be mashed and stir-fried slowly on medium-low heat until it bubbles.

米粉如何炸至鬆脆？
米粉放入滾油後，待米粉炸起及香脆即可，動作敏捷，時間毋須太久，否則米粉變硬焦燶。

How to make deep-fried rice vermicelli crunchy?

Put the rice vermicelli in hot oil. When they swell and turn crisp, it is done. Dish up quickly; otherwise they will turn hard and scorch.

黃金豆腐醬可伴甚麼享用？
黃金豆腐醬豆香味濃，伴飯、意粉、多士或餅乾作為派對小吃，同樣感受超凡的味覺享受！

What food can also be served with the salted egg yolk and tofu paste?

The paste of rich soybean flavour can also be served with rice, spaghetti, toast or biscuits as the party snacks. All will give you the same extraordinary taste!

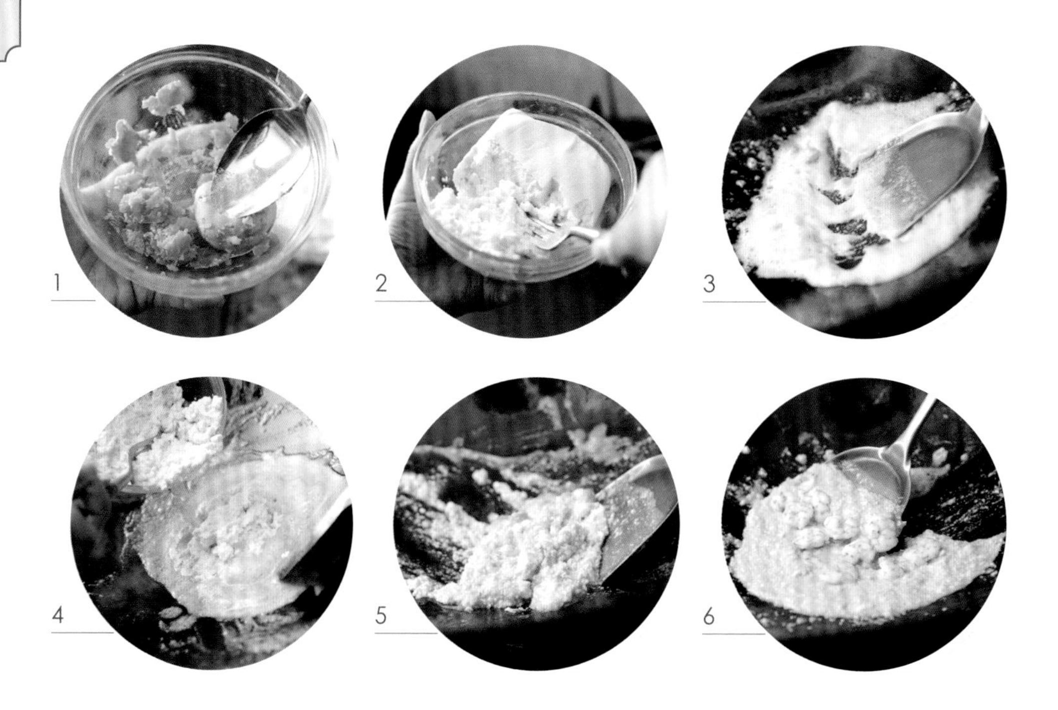

Ingredients (Serves 4)

1 piece tofu
150 g shelled shrimps
2 salted eggs
1 serving rice vermicelli

Seasoning

1/3 tsp sugar
1/2 tsp salt
1 tsp cornflour
2 tsp sesame oil
3 tbsp water

Preparation

1. Rinse the whole salted eggs. Put in water and heat until it boils. Blanch for about 6 minutes until done. Mash the egg yolks. (picture 1)
2. Rinse and scald the shrimps. Crush the tofu with a fork. (picture 2)

Method

1. Put the rice vermicelli in hot oil. Deep-fry for a moment until crisp. Drain and set aside.
2. Heat a wok. Put in 2 tbsp of oil. Stir fry the egg yolk until it is aromatic and bubbles. Add the tofu and shrimps. Cook for a while. Put in the seasoning and bring to the boil. Mix well. Pour on top of the rice vermicelli. Mix well and serve. (pictures 3-6)

芝士焦糖脆蛋

Cheese Crème Brûlée

材料（可製成 4 杯）

忌廉芝士 120 克
忌廉 2 杯
糖 4 湯匙
蛋黃 4 個

脆蛋面

糖 6 湯匙

藍莓醬

藍莓 1 盒
檸檬汁 2 茶匙
糖 3 湯匙
水 1/3 杯
粟粉 2 茶匙
（與水 1 湯匙調勻）

預先準備

1. 忌廉芝士與忌廉半杯拌勻（可用電動打蛋器打滑）。
2. 蛋黃與糖隔熱水打至質感濃厚。
3. 焗盤內注入熱水。焗爐調至 180℃，預熱 10 分鐘。
4. 藍莓醬做法：藍莓加入水、糖、檸檬汁煮片刻，注入粟粉水煮至濃稠，待涼。

做法

1. 煮熱忌廉 1.5 杯，加入已拌勻的忌廉芝士拌勻，以慢火煮片刻。
2. 慢慢逐少加入蛋黃混合料內，拌勻。
3. 將四個焗盅放在已注入熱水的焗盤內，舀進忌廉芝士混合料，以烘焙紙蓋着。
4. 放入已預熱的焗爐焗 30 分鐘，取出，待涼後雪凍。（圖 1-2）
5. 每個蛋面灑上 1.5 茶匙糖，用火槍燒焦糖面，伴藍莓醬進食。

零失敗技巧
Successful cooking skills

為甚麼蛋黃與糖要隔熱水才用打蛋器打透？
這樣才容易將蛋黃與糖打至濃厚。

Why do you beat the egg yolks and sugar over a pot of simmering water?
This helps whip in more air in to the custard making it fluffy as the egg yolks are partly cooked.

用哪款火槍？
附圖的火槍皆可購自烘焙材料店，但無論哪款都要小心使用。

Which kind of torch should I use?
The torches shown in the photo on this page are all available from baking supply stores. No matter which one you choose, use it carefully.

1

2

Custard (makes 4 servings)

120 g cream cheese
2 cups whipping cream
4 tbsp sugar
4 egg yolks

Caramel topping

6 tbsp sugar

Blueberry sauce

1 pack blueberries
2 tsp lemon juice
3 tbsp sugar
1/3 cup water
2 tsp cornflour (mixed with 1 tbsp of water)

Preparation

1. Beat cream cheese with 1/2 cup of whipping cream until smooth with an electric mixer.
2. Mix egg yolks and sugar in a bowl. Heat over a pot of simmering water while beating with an electric mixer until thick and dense.
3. Pour boiling water into a deep baking tray. Preheat an oven to 180˚C for 10 minutes.
4. To make the blueberry sauce, cook blueberries in water, sugar and lemon juice for a while. Stir in cornflour solution and cook until thick. Leave it to cool.

Method

1. Heat 1.5 cups of whipping cream until hot. Add whipped cream cheese into the hot cream. Stir well. Cook over low heat for a while.
2. Slowly stir the cream cheese mixture into the egg yolk mixture. Stir well.
3. Put 4 ramekins into the deep baking tray with boiling water. Divide the cream cheese mixture among the ramekins. Cover with baking paper.
4. Bake in a preheated oven for 30 minutes. Leave it to cool at room temperature. Then refrigerate until cold.
5. Sprinkle 1.5 tsp of sugar on each serving of custard. Burn the sugar with a kitchen torch until lightly browned and runny. Serve with blueberry sauce on the side. (pictures1-2)

豆腐西米焗布甸

Baked Tofu and Sago Pudding

材料（4 人份量）

盒裝豆腐 330 克
西米 3 湯匙
雞蛋 2 個
栗子茸 4 湯匙
粟粉、吉士粉各 2 湯匙
牛油 1 片
砂糖 3 湯匙
糖霜 1 湯匙

預先準備

1. 預熱焗爐至 200℃。
2. 西米用過面清水浸 1 小時，隔去水分。放入滾水內，用小火煲約 5 分鐘至西米呈透明，過冷河，瀝乾水分備用。

做法

1. 豆腐用叉子壓碎；粟粉、吉士粉、雞蛋及水 1/4 杯拌勻成蛋漿。
2. 牛油、砂糖及水 1 杯拌勻，煮至糖溶化，加入豆腐碎煮滾，再傾入西米（餘下少許西米）及蛋漿用小火煮成糊狀。（圖 1）
3. 豆腐糊漿傾入小焗盅至半滿，加栗子茸 1 湯匙，再鋪入豆腐漿至滿，在表面鋪上西米及灑上糖霜，排於盛有水的焗盤，焗約 10 分鐘至表面呈焦黃色即成。（圖 2-4）

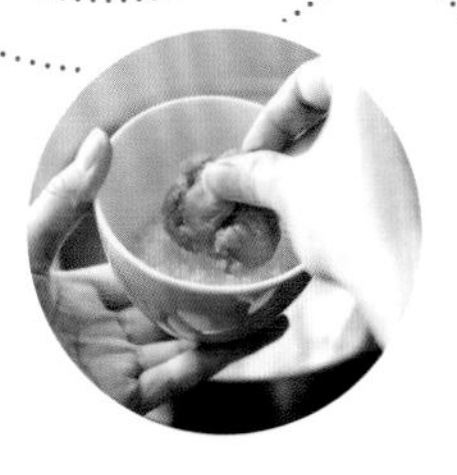

1 2 3 4

Ingredients (Serves 4)

330 g packed tofu
3 tbsp sago
2 eggs
4 tbsp chestnut puree
2 tbsp cornflour
2 tbsp custard powder
1 slice butter
3 tbsp granulated sugar
1 tbsp icing sugar

Preparation

1. Preheat an oven to 200°C.
2. Soak the sago in water for 1 hour (water covering all the sago). Drain well. Put the sago in boiling water. Cook over low heat for about 5 minutes until transparent. Cool in cold water. Drain and set aside.

Method

1. Crush the tofu with a fork. Mix the cornflour, custard powder and eggs with 1/4 cup of water.
2. Mix the butter, granulated sugar and 1 cup of water evenly. Cook until the sugar melts. Add the tofu and bring to the boil. Stir in the sago (remains a little of sago for later use) and egg mixture from step (1) . Cook over low heat until thick. (picture 1)
3. Put the tofu paste into a small baking dish in half full. Add 1 tbsp of the chestnut puree. Top with the tofu paste until full. Sprinkle the icing sugar on the surface. Arrange on a baking tray containing water. Bake for about 10 minutes until the surface turns golden brown. Serve. (pictures 2-4)

零失敗技巧

Successful cooking skills

為甚麼將焗盅排於盛有水的焗盤烘焗？
以免西米焗布甸太乾太硬，欠軟滑口感。

Why arrange the baking dish on a baking tray containing water to bake?
Baking with water is to avoid the pudding becoming too dry and firm, losing its soft and smooth taste.

豆腐味濃郁嗎？
豆香與蛋味混和，是西米焗布甸之新口味，賓客一定喜歡！

Does it have a strong beancurd flavour?
It has a rich flavour of soybean and egg – another choice of sago pudding. The guests must be love it.

編者 Editor
Forms Kitchen編輯委員會 Editorial Committee, Forms Kitchen

美術設計 Design
羅穎思 Venus Lo

排版 Typography
何秋雲 Sonia Ho

出版者 Publisher
Forms Kitchen

香港鰂魚涌英皇道1065號
東達中心1305室
Room 1305, Eastern Centre, 1065 King's Road,
Quarry Bay, Hong Kong.
電話 Tel: 2564 7511
傳真 Fax: 2565 5539
電郵 Email: info@wanlibk.com
網址 Web Site: http://www.wanlibk.com
http://www.facebook.com/wanlibk

發行者 Distributor
香港聯合書刊物流有限公司 SUP Publishing Logistics (HK) Ltd.
香港新界大埔汀麗路36號
中華商務印刷大廈3字樓
3/F., C&C Building, 36 Ting Lai Road,
Tai Po, N.T., Hong Kong
電話 Tel: 2150 2100
傳真 Fax: 2407 3062
電郵 Email: info@suplogistics.com.hk

承印者 Printer
中華商務彩色印刷有限公司 C & C Offset Printing Co., Ltd.

出版日期 Publishing Date
二零一九年一月第一次印刷 First print in January 2019

Published in Hong Kong by Forms Kitchen,
a division of Wan Li Book Company Limited.
ISBN 978-962-14-6961-8

鳴謝以下作者提供食譜（排名不分先後）：
黃美鳳、Feliz Chan、Winnie姐